THE BODY MAP SERIES
The BODY MAP
The BODY MAP
The BODY MAP
The BODY MAP
The BODY MAP

The Body Map: Designed From The Beginning *Guide to the Homo sapien*

2016-2025 Worlds of Wonder Publishing
Dallas, Texas

ISBN-13: 978-0-9966317-1-6

Cover Design- Tim Cocklin

Address all correspondence to:
tim@worldsofwonderpublishing.com

Print copies are available from Amazon. Kindle versions are also available at a reduced cost. If you can not afford a copy contact me at the email address tim@timcocklin.com.

This book is dedicated to:

The Creator of all things, his thoughts and ways are far beyond comprehension.
I pray this book is pleasing to him.

To Mr. McGrew my Sunday school teacher who told us "If you don't remember anything else I tell you, remember: **In the beginning God**"

To Bill and Shirley Gay who took a young stranger off the street and treated me like a son.

To Ken Kercher a very true and dear friend who gave a young man opportunities to fail and grow.

To my children Petrushka and Aaron who stuck with me and prayed for me through some very dark times.

To my wife Cindy who stayed when she should have left and would not let me die but continued to support and encourage me to write this book, letting me spend endless hours researching the subject.

To the reader, it is my hope that this series of books will generate a lot of questions in your mind about evolution and the idea that life is nothing more than a series of serendipitous accidents and undirected chance.

OTHER BOOKS - By Tim Cocklin
https://www.worldsofwonderpublishing.com

History and Operation of Believers in the 1st century, in faith

Guide to Passover for Christians with The

A Study of Leviticus 23 The feasts their

Hidden details of the end-times

The testimony and life of Bill and Shirley Gay

Details about the who, why, and how of reversing metabolic

Reprint - Public Domain - The Coming Prince - 1894

Reprint - Public Domain Heavenly Dream

God's divine intervention in history

The modern Church is consumed w/ apostates What

Beginning Hebrew Learn Biblical Hebrew **Coming Soon -**

Beginning Greek Learn Biblical Koine Greek **Coming Soon -**

OTHER BOOKS - By Tim Cocklin
https://www.worldsofwonderpublishing.com

I
How the body was designed from a microbiological

II
The miracle of reproduction in humans.

III
How the human body operates: The Five Senses

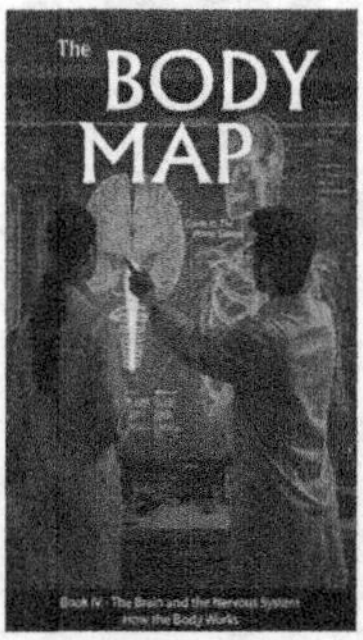

IV
How the human body operates: The Brain

How the human body operates: Organs of the Body
COMING SOON

How the human body operates: Endocrine System

How the human body operates: Circulatory System & The Heart

How the human body operates: The Lymphatic System

How the human body operates: Respiratory system
COMING SOON

How the human body operates: The Digestive System
COMING SOON

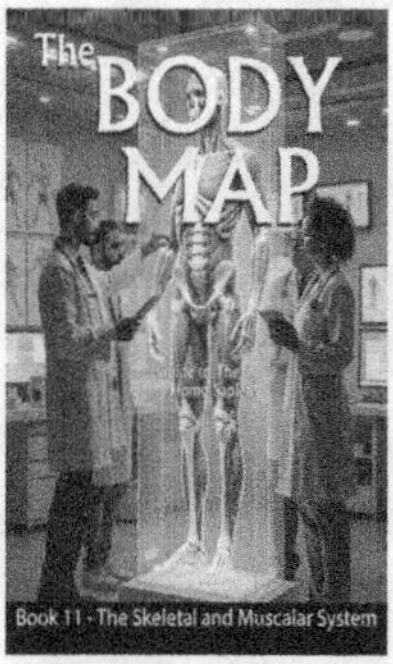

How the human body operates: Skeletal Muscular systems
COMING SOON

How the human body works All Systems Step-by-step
COMING SOON

Designed from The Beginning

Wishful thinking does not create change.

Science programs often tell us that creatures like the box jellyfish evolved their deadly venom for self-defense. While this explanation sounds reasonable, it assumes that biological change is driven purely by need—a claim that doesn't hold up under scrutiny. Just because an organism needs protection doesn't mean it can rewrite its DNA to produce a complex toxin. That capability would have to exist from the start. Otherwise, the creature wouldn't survive the countless failures it would face trying to develop such a trait. Even if one individual somehow did, there's no guarantee the change would pass to its offspring.

This raises a critical question for advocates of evolution: how does a system purposefully alter its own DNA? The standard answer is mutations. But observational science tells a different story. DNA remains largely stable over generations—a phenomenon known as stasis. When mutations do occur, they are typically neutral or harmful, not beneficial. We've never observed a dog giving birth to anything but another dog. The limits of the gene pool seem fixed, bound by the constraints of its original design.

The idea that random mutations can generate entirely new species or permanent, beneficial traits is not a conclusion based on evidence—it's a belief. Evolutionary theorists often use language like "evolutionary pressures caused" certain traits to emerge, which suggests purpose. But evolution, as defined by its own theory, is a purposeless, undirected process. Chemicals don't have goals or intentions.

Then there's the biggest unanswered question of all: how did life begin? Evolution doesn't address the origin of life—how living matter could come from non-living material. Nor does it explain the astonishing complexity of the human body, such as the precise coordination of hormones, as the product of chance. Claiming this complexity arose accidentally seems far more miraculous than proposing an intelligent designer.

Ultimately, this isn't just a scientific issue—it's a philosophical one. Many who cling to evolutionary theory do so out of a deep commitment to naturalism and humanism. In rejecting the idea of design, they adopt a belief system of their own—one rooted not in conclusive evidence, but in unwavering faith in material causes alone.

– Tim Cocklin

TABLE OF CONTENTS

CONTENTS

When a legitimate criticism of Darwinism is presented, it is often dismissed without a sufficient response—brushed aside on a technicality, countered with an unrelated argument, or simply ignored altogether. Over time, the public tends to forget that these criticisms were never properly addressed by Darwinists. Nevertheless, Darwinism continues to dominate the scientific narrative. Because these critiques failed to displace Darwinism from its position of influence, many assume—incorrectly—that the criticisms must have been thoroughly discredited or refuted at some point. (William Dembski, *Uncommon Dissent: Intellectuals Who Find Darwinism Unconvincing*, p. xxv)

I highly recommend his book to those seeking a more thoughtful and intellectually grounded critique of Darwinism. The rejection of Intelligent Design and Creation Science is not based on the observable evidence of scientific discovery, but rather on a philosophical bias—a rigid commitment to naturalism that excludes any consideration of design or purpose.

It's important to note, however, that advocates of Intelligent Design do not necessarily endorse a six-day creation or affirm the Christian God as the Creator. Nevertheless, they strongly reject the notion that blind chance and natural selection can fully account for the complexity and order we observe in the natural world—and on that point, we find common ground.

This kind of selective reasoning is not limited to the scientific community. We see a similar pattern in certain sectors of the media, where conservative viewpoints are often dismissed—not on the basis of truth, but due to ideological bias. Evidence that doesn't align with a predetermined narrative is simply ignored. For instance, radical ideologies are often excused or downplayed, while Christians and conservatives are unfairly portrayed as extremists. This inversion of truth is deeply troubling.

We are living in a time when common sense is under siege and falsehoods are widely accepted as fact. I call it "the great exchange"—the tragic replacement of truth with lies.

> **"...It does not require a majority to prevail, but rather an irate, tireless minority keen to set brush fires in people's minds.."** -Samuel Adams

Yes, we are entering very dark times. I believe things will get even worse, the Bible has predicted it. Please prepare your heart for the coming malaise! Make sure of your relationship with the Creator of all things. This is not our home, we are looking forward to spending eternity with our Creator.

> *This know also, that in the last days perilous times shall come. For men shall be lovers of their own selves, covetous, boasters, proud, blasphemers, disobedient to parents, unthankful, unholy, Without natural affection, trucebreakers, false accusers, incontinent, fierce, despisers of those that are good, Traitors, heady, highminded, lovers of pleasures more than lovers of God;*
>
> *- 2 Timothy 3:1-4 KJV*

When you see a word with this ♥ after it, it indicates that the word is defined in the glossary. When you see a word with § that subject will be found in appendix D special subjects.

Propaganda in Schools

Not Just a Theory

Next time someone tries to tell you that evolution is just a theory, as a way of dismissing it, as if it's just something someone guessed at, remember that they're using the non-scientific meaning of the word. If that person is a teacher, or minister, or some other figure of authority, they should know better. In fact, they probably do, and are trying to mislead you.

The above image was a part of a homework assignment page from an 8th-grade class at Heritage Middle School in Deltona, Florida.

Not just a Theory

> **"Next time someone tries to tell you that evolution is just a theory, as a way of dismissing it, as if it's just something someone guessed at, remember that they're using non-scientific meaning of the word. If that person is a teacher, or minister, or some other figure of authority, they should know better, In fact, they probably do, and are trying to mislead you."**

Heritage Middle School received so many complaints that it had to retract the lesson and issue an apology letter. It shows how far the evolutionist goes to promote the idea of evolution, to go so far as to say any authority figure that tells you otherwise, is a liar.

In fact, we will show in this book, that not only is Darwin's so-called idea not a theory, it isn't even a good hypothesis from a scientific point of view using the well known and respected "scientific method". Many Ph.D level scientists, after examining the idea carefully, have concluded that it could never happen, even those espousing a belief in evolution.

Official science: The grand illusion

Posted on January 15, 2016 - PersonalLiberty.com

"Government science exists because it is a fine weapon to use, in order to force an agenda of control over the population. We aren't talking about knowledge here. Knowledge is irrelevant. What counts is: 'How can we fabricate something that looks like the truth?' I keep pointing this out: we're dealing with reality builders. In this case, they make their roads and fences out of data, and they massage and invent the data out of thin air to suit their purposes. After all, they also invent money out of thin air."

— *"The Underground"* by Jon Rappoport Http://www.nomorefakenews.com/

Introduction

Before we start, I want to alert you to the fact that I will use a lot of terms in this book which will most likely not be familiar to you. Additionally, I quote scientists with a Ph.D level of education and their language can be overwhelming and hard to understand at times. It is not my intent to be pedantic*, therefore I have included in Appendix C a glossary of words♥ and their meanings. Please check the glossary if a word is not known to you. I hope it helps you understand the point being made.

* Characterized by a narrow, often ostentatious concern for academic knowledge and formal rules. - www.thefreedictionary.com

> **"Darwin's theory of evolution is the great white elephant of contemporary thought. It is large, almost completely useless, and the object of superstitious awe."**
>
> *Dr. David*

This series of books is a map of the human body with all of its intricate parts. Your worldview will determine how you view life's beginnings. If you believe there is no God, then you will follow the path of evolution and try to prove there is no God or designer. If you believe there is a God, then you will follow some form of creation. In either case, it takes faith. However, the Bible is the only book that accurately predicts future events, that alone should prove the validity of the book! Neither belief is based on first-hand observation, neither can be repeated. We must then base our assumptions on what we can see now. Darwin said:

> "If it could be demonstrated that any complex organ existed, which could not possibly have been formed by numerous, successive, slight modifications, my theory would absolutely break down."

Our attempt here is to show that not only could the body never have developed by small steps, but it is also designed and programmed in a way that undirected chance could never have been the cause of all we see. What Darwin did not know staggers the mind. With all of the advances in microbiology, there are those who still willingly refuse to accept the obvious truth because of their bias against the Creator or a Designer.

It is not about evidence, it is about your Interpretation of the evidence.

The human body is highly programmed, unlike anything that man has ever been able to develop. I spent the last 30 plus years programming, starting with FORTRAN then PL/1 and then to C and finally Websites using PHP and mySQL and now AI. There are a lot of similarities between how the body works and how a computer operates. Both have an operating system, both have feedback loops and "what if scenarios". However, the body can heal its hardware. As of this writing, I do not know of self-healing computer hardware. There are software products that can heal their processes, but none that can heal a chip that has gone bad without shutting down the whole system and replacing the part(s), unless they are hot-swappable.

Each human cell—of which the body contains trillions—is an incredibly advanced information-processing system. It includes numerous interdependent components that function like multiple operating systems and programming languages, complete with mechanisms for encoding and decoding, communication, error correction, and input/output controls for cellular operations. These complex features work together to carry out life's essential processes. According to Dr. Donald E. Johnson, the real difficulty for any naturalistic explanation of life's origin lies in explaining how such a system could arise without intelligent guidance. The highly ordered digital information in living cells mirrors, and even surpasses, man-made computer systems, yet natural laws—such as those of chemistry and physics—are insufficient to generate or organize this kind of functional, algorithm-driven information on their own.
Probability's Nature & Natures Probability *-A Call to Scientific Integrity* (p. 47)

To believe that everything we see was the result of a series of serendipitous chance happenings, then made permanent by natural selection, is tantamount to saying that Bill Gates mixed a series of random bits (1s and 0s) and stirred them up until Microsoft Notebook was created. Then using that as a base, pouring in more bits, and Microsoft Word was born. The human body is billions of times more complex and functional, able to self-perpetuate, fix issues on the fly, heal itself and initiate a thought among many other things.

> Human DNA is like a computer program but far, far more advanced than any software we've ever created.
>
> Bill Gates, The Road Ahead, p.228.

The human body has an operating system that runs on a complex set of algorithms and subroutines that dictate how and when things happen. For instance, how would the body, through undirected evolution, know that we need baby teeth when we are developing, because of the size of our jaw and head and adult teeth later? Having adult teeth to start would cause the jaw to break or at the very least, cause massive deformation. How would the body know when to start releasing eggs in the female at a certain age? If a very young child were to get pregnant earlier, their body would not be able to handle the fetus forming in its body. What determines when the body should stop growing, when we reach a certain age, why not keep growing throughout life? Undirected chance could NEVER predict the necessity for this, it is blind to the consequences of its so-called fortuitous "mutations". It takes a designer to put all of those things in place, and to design a system that stops growing at a certain point. I will attempt to answer all of those questions through out the 12 volumes of the Body Map series.

In the development stage, there is a process called apoptosis. Apoptosis is programmed cell death where cells are killed on time, on purpose. Without it you would have webbed hands and feet because the process kills the cells forming fingers and toes. After development, apoptosis is a "programmed" process whereby cells die of natural causes. When apoptosis is not able to carry out its normal process "cancerous" cells form and grow uncontrollably.

Cancer develops when cells begin to grow uncontrollably and fail to die as they should. In a healthy body, cells grow, divide, and eventually undergo a natural process of self-destruction called apoptosis. When this process is disrupted, cells that should die continue to survive and multiply. Unlike normal cells, cancer cells bypass this built-in death program, allowing them to keep dividing and forming a growing mass of abnormal tissue. (Paraphrased from "What is Cancer?" by Peter Crosta, Medical News Today, September 2008 © MediLexicon International Ltd)

The problem with computer software is that it needs hardware to be functional. The problem with the body is that it has to have an operating system to function. DNA alone will not suffice. There has to be something to direct the body to function in very specific ways. The body's operating system has to keep us breathing without our being conscious that we are breathing. (called the Autonomic Nervous System - Appendix D §). If we get a cut on our finger, the body has to stop the bleeding and then heal the wound without clotting all the blood in our body. The brain has to be able to filter out noises, or we would be distracted all the time. The brain will also filter out smells, or we would continue to be overwhelmed (see book 3, The Five Senses). There are thousands of these functions that we take for granted since they are automatic. We could go on and on listing functions. We will take each part of the body and attempt to study it in minute detail, describing what happens in each system and why it is virtually an "all or nothing system.""

While doing the research for the book it came to me like a lightning bolt.

Undirected chance mutations *can never* anticipate future problems that may arise in its environment!

Natural Selection takes place after the fact so it is no help. **ONLY** a designer would know how to build in contingencies! Without these built-in contingencies the organism would die and that would end the process of evolution. One example is the blood clotting cascade. How could undirected chance mutations anticipate the need to stop blood flow and only at the source of the wound? Trial and error in this very complex process will most certainly eliminate the organism immediately upon the first cut!

-Tim

"What Darwin did not know now fills volumes of books. We will endeavor, in this book, to show that evolution is an outdated concept based on faulty logic and unscientific methods. What if what Darwin postulated had been first introduced today? With all that we now know, he would have been laughed out of the scientific community. Darwin knew nothing of DNA or even what happens inside a cell. He had no ability or access to what we know in microbiology. What the evolutionist attempts to do is make their point with "matter of fact" statements not backed up by any explanation. For instance, why do we have two eyes, two ears, two arms, and two legs? How would undirected chance have created a perfectly tuned man or woman with all of the feedback systems and a working immune system that protects us from disease? Chance does not explain it, and natural selection and survival of the fittest only occur when a function exists. Yes, Darwin's theory is broken down because, in too many cases, physical functions could never have come about by "numerous, successive, slight modifications.""

> ***"The first gulp from the glass of natural sciences will turn you into an atheist, but at the bottom of the glass God is waiting for you."***
>
> **Werner Heisenberg**
> **Nobel laureate in physics**

This series will show that there is clear evidence that life critical systems, could not survive successive steps. Michael Behe coined the term "Irreducibly Complex" when referring to the bacterial flagellum. We can call the human machine virtually irreducibly complex because it is "ALL-OR-NOTHING" in most systems in the body. There are systems in the human machine that will cease to functioning if all parts are not working or present. The point of these books is to show just how complex and vulnerable we are as a species. Life is very delicate. God, in His wisdom, built into us protection mechanisms which will fight off intruders and repair facilities that will fix problems that arise as we go through life. Evolution and natural selection could never look forward and anticipate problems that may arise, that takes a designer. You will learn that the human machine is like a finely tuned computer program. That is why this first book is called Designed From The Beginning.

Your body is a finely tuned program. From fertilization to death, your body is running what I will call, for the lack of a better term, a computer program. There is intelligence built in that can never be explained by evolution or natural selection. First, what intelligence directs the cells to form different body parts in a growing fetus, and in what order? (See Cellular Differentiation in Embryonic Development - Appendix D §) Second, what triggers the birth process? Third, how does the body know when to stop growing? Fourth, how does the body know how and when to protect itself? These are just a few questions that will prove that we were designed from the beginning.

The body was created to operate in a particular way. It is when we stress it and add foreign elements that the body begins to break down. We know, for instance, that lack of sleep, being overly stressed, and living a sedentary lifestyle all have an effect on the smooth operation of the machine. In our modern world, here in the United States, we have plenty of food, but the quality of that food has been greatly reduced by over-processing, added sugar, and trans fats. Add that to the increase of GMOs (genetically modified organisms), and we get an increase in multiple diseases. Ask yourself: why is there an increase in type 2 diabetes among adults? Why is heart disease the number one killer? You will discover that chemicals added to everyday items and food cause many of the problems."Homeostasis, we discuss Endocrine Disruption which is likely the cause of many of the illnesses and problems we experience today.

> "A Lie doesn't become truth, wrong doesn't become right and evil doesn't become good, just because it's accepted by a majority."
>
> Booker T. Washington

A process that is now being investigated in more detail, known as Epigenetics, explains a lot about why we are not only the product of our DNA but are also affected by what we eat, how we live, and environmental factors. We will discuss Epigenetics in more detail in the chapter on DNA.

The Beginning

After an introduction to microbiology and the cell, we will start the series with reproduction in Book II. That is the beginning for each one of us. Life begins at conception, not when the child comes out of the mother's birth canal and breathes its first breath. Without this process, we would not exist. The number of systems that must be in place for a new person to be created is staggering! You will also find that the growing baby is not just a passive bystander. I will continue to say it is the "all-or-nothing" proof of creation. Life could never have continued without the reproduction process, and reproduction cannot occur without all of the systems and programming in place to make it happen. You will find that the body not only has an incredible design, but is a finely tuned program, not unlike a computer program. The body is controlled by a set of instructions and commands. It is the most complex program on earth. When you get a cut on your finger, the program that executes the process of healing is totally amazing. Evolution and natural selection could never account for the amazing intelligence that takes place in an instant.

> In the final 1876 printing of the 1872 sixth edition, Darwin employed the word "may" 642 times, "if" 493 times, "might" 203 times, "probable" or "probably" 182 times, "tend" or "tendency" 153 times, "suppose(d)" 141 times, "perhaps" 63 times, "no doubt" 58 times, and so on. Exploring the God Question 2. Life and Evolution, Part 1 (Darwinism) exploring-god-question by Russell Grigg Published: 13 January 2015

Have you ever thought about how you stop bleeding after a minor cut? If the body did not have a mechanism for stopping blood flow, you would die in a fairly short time. If the coagulation process did not stop at the wound site, it might coagulate all the blood in your body once the process started, and you would die within minutes. However, the Great Designer has created a fascinating system for handling this problem.

Blood clotting is a multi-step process with many dependencies. The first step is vascular constriction, limiting blood flow to the injured area. Next, platelets become activated by thrombin, a protein found in the bloodstream. This forms a temporary platelet plug. The protein fibrinogen is responsible for platelet clumping. The platelets clump by binding to collagen, which was exposed during the cut in the lining of the vessels. When activated, the platelets release ADP and thromboxane A2, a lipid known as an eicosanoid. Both of these release additional platelets, serotonin, phospholipids, lipoproteins, and other proteins critical to the coagulation cascade. The platelets then change their shape to help form the plug. Then, the platelets are meshed in a network of fibers known as a fibrin mesh, and the clot is formed. Finally, after a few days, plasmin begins to dissolve the clot and repairs it completely. We will discuss this in much more detail in Book VII, Circulatory System, and the Heart.

> "Unfortunately,in the field of evolution most explanations are not good. As a matter of fact, they hardly qualify as explanations at all; they are suggestions, hunches, pipe dreams, hardly worthy of being called hypotheses.."
>
> Norman Macbeth, *Darwin Retried* (1971), p. 147.

The human body is made of many complex interrelated systems. These systems can communicate and somehow initiate the processes automatically to protect the body from significant irreversible damage.

Think, for just a moment, what part(s) of your body you could live without and still survive. You can live without sight (but not without someone who can see). You can live without hearing (but you would not hear the impending danger). You can live without taste (although you may eat the wrong thing). You can live without feeling (but you may not realize you are injured and die). All of those outward things aside, can you live without a heart? Can you live without a liver, kidneys, lungs, a brain, or any of the other organs? At a lower level, could you survive without red and white blood cells? More detail can be found in Book III, The Five Senses.

We could keep going, but the point is that the human body is an integrated whole. When any of the multitudes of systems break down, we have trouble. So, where do the evolutionists get away with telling us that we were made by small incremental steps, by mutations, and then maintained by natural selection? Evolution is a religion, a philosophy, an unsubstantiated idea, NOT science. Once God breathed into man the breath of life, His creation was finished. He told man to fill the earth. Man now had everything in him to perpetuate the race. And life continues unaided by evolution or any other nonsense!

> Responding to Pat Robertson's comments that Young Earth Creationists are "deaf, dumb and blind," Ken Ham of the Creation Museum accused the CBN host of compromising Scripture and thus serving as "one of the biggest problems we have today in the church."
> ***I agree!***

The Discovery Channel, National Geographic, Nova, and other "science" media outlets do not go deep enough to show the true complexity of life. This may be intentional. If everything seems simple, then evolution can be perceived as believable, and when you add billions of years, it seems plausible. It is when you begin to see the incredible complexity, the critical interdependencies, and design that you realize that **NO** undirected natural process could ever anticipate all of the necessary feedback loops and future contingencies that are required for life! If this series piques your curiosity and leaves you with questions that cause you to do the research to come to your conclusion, then the book is a success.

You are free to believe anything you want, but you can only make a truly free choice if you know both sides of an argument. In most cases, because of the indoctrination of evolution in the public school system, you have no choice but to believe what you are told. This is why many good scientists are blinded by an ingrained bias towards evolution. Remember the assignment given to the 8th grade in Florida? It is not just: this is what we think; it is this is what you will think.

The best way to learn anything is to ask questions (watch any two-year-old who begins to talk). In my research, I have discovered that many strict adherents to the so-called evolution theory have rejected the hypothesis after students they were teaching asked probing questions about flaws in Darwinism. This is why people like the Executive Director of the National Center for Science Education (at the time of this writing) must demand that school boards make sure that evolution is never questioned in the classroom. It is critical that the flaws in the hypothesis never be discussed.

Once these former sycophants (those who flatter the evolutionist by willingly accepting their belief) began to explore the presuppositions and the evidence, it became clear that evolution could never explain the origin of life and certainly not the far-fetched, evidence-free belief of macro-evolution.

> "There is no chance (<10^{-1000}) to see this mechanism appear spontaneously and if it did, even less for it to remain."
>
> - Marcel P. Schutzenberger
> Algorithms and the Neo-Darwinian Theory of Evolution,

This is tantamount to book burning in Nazi Germany; freedom of thought cannot be allowed if you are going to control the population.

You must ask yourself why the educational system will not allow a comparison of Creation or, at the very least, Intelligent Design. Intelligent Design does not require a belief in the God of the Bible, only a belief that life was designed by some entity. This will fit nicely with the theory of Directed Panspermia, which we will discuss later.

Dr. Dean Kenyon co-authored a book called Biochemical Predestination. In this book, he put forth the idea that biochemical molecular compounds, such as proteins, could assemble themselves from non-living raw chemicals given the right environmental conditions. This book is still used in universities today to prove evolution. However, Dr. Kenyon recanted his hypothesis because the theory of biochemical predestination fails to explain how complex amino acid-based proteins self-assemble without or before the existence of DNA-based sequencing or assembly codes. Dr. Kenyon is a rare example of clear, unbiased thinking. To admit that his hypothesis was incorrect after gaining such notoriety is truly the mark of a great man.

> "Darwin's theory, I believe, is on the verge of collapse. In his famous book, Origin of Species, Darwin made a mistake sufficiently serious to undermine his theory. And that mistake has only recently been recognized as such. I have not been surprised to read ... that in some of the latest evolutionary theories 'natural selection plays no role at all.' Darwin, I suggest, is in the process of being discarded ... But perhaps in deference to the venerable old gentleman ... it is being done as discreetly and gently as possible, with a minimum of publicity"
>
> Tom Bethell, "Darwin's Mistake,"
> The Craft of Prose, pp. 311, 314,

If you do not know the facts behind why the God of the Bible created everything, you can easily fall into the evolutionist's trap. The evolutionist is very adept at making the illogical appear logical. By twisting and spinning a yarn that even Mark Twain would be proud to have authored, they convince the uninformed that natural cause is the only truth and Creation is just superstitious nonsense and not scientific. As a result, many have lost their faith. It is, therefore, critically important that you, as a Christian, have the facts on hand to refute the many evolutionary incongruences.

> *But sanctify the Lord God in your hearts: and be ready always to give an answer to every man that asketh you a reason of the hope that is in you with meekness and fear: 1 Peter 3:15 (KJV)*

Bob Dylan wrote a song "You Have Got To Serve Somebody" you either serve the God of Creation or the Prince and the power of the air (Satan). That may sound self-serving and arrogant but I believe it with my whole heart. Some may say that we do not serve anyone but ourselves, there is no God confirming what Psalm 14 says.

> *Psalm 14:1 The fool hath said in his heart, [There is] no God.*

We make our own way in life. We are our own god. But in the end, it is clear, you are not your own god, as we hope to prove in this book.

How big is our God? Try to imagine how big or intelligent God is; whatever you imagine would fall very short of His greatness. We do not have the capacity to understand God in His fullness, but He has given us clues. The more we learn about the universe, and everything down to our DNA, the more we must realize how much we do not understand. God has made it possible for us, in this day of great scientific discovery, to understand some of the greatness of His creation.

> *For the invisible things of him from the creation of the world are clearly seen, being understood by the things that are made, even his eternal power and Godhead; so that they are without excuse: Romans 1:20 (KJV)*

Anyone who studies this creation must, if they are honest, acknowledge that an intelligent source is behind everything we see. However, there are those who reject this idea because they reject God. They have the most ridiculous theories and ridicule those who do not follow their theories blindly. Even Francis Crick, co-discoverer of the structure of DNA, realized that chance could never have created the detailed information in DNA and put forth the idea that aliens seeded the world with this amazing information (Directed Panspermia). It shows how far people will go to reject the Creator God.

> An honest man, armed with all the knowledge available to us now, could only state, that in some sense, the origin of life appears to be almost a miracle, so many are the conditions which would have had to

have been satisfied to get it going. - Francis Crick, *Life Itself -- Its Origin and Nature, Future,* 1982.

Peter calls these types willingly ignorant.

> *Knowing this first, that there shall come in the last days scoffers, walking after their own lusts, And saying, Where is the promise of his coming? for since the fathers fell asleep, all things continue as they were from the beginning of the creation.* ***For this they willingly are ignorant*** *of, that by the word of God the heavens were of old, and the earth standing out of the water and in the water: Whereby the world that then was, being overflowed with water, perished: But the heavens and the earth, which are now, by the same word are kept in store, reserved unto fire against the day of judgment and perdition of ungodly men. 2 Peter 3:3-7 (KJV)*

They know that the beliefs they hold are feeble, at best, and yet they continue to promote these ideas in the classroom, on TV, and in books. Because if they allow anything but natural processes to explain the beginning of life, they become accountable to God. An example of this contempt comes from evolutionists Richard C. Lewontin and Thomas Nagel:

> "It is not that the methods and institutions of science somehow compel us to accept a material explanation of the phenomenal world, but on the contrary, that we are forced by our a priori* adherence to material causes to create an apparatus of investigation and a set of concepts that produce material explanations, no matter how counter-intuitive, no matter how mystifying to the uninitiated. Moreover, that materialism is absolute, ***for we cannot allow a Divine Foot in the door."*** [my emphasis]

*Definition of a priori from the Web: involving deductive reasoning from a general principle to a necessary effect; not supported by fact; "as a priori judgment"

> I want atheism to be true and am made uneasy by the fact that some of the most intelligent and well-informed people I know are religious believers. It isn't just that I don't believe in God and, naturally, hope that I'm right in my belief. It's that I hope there is no God! I don't want there to be a God; I don't want the universe to be like that. - Thomas Nagel, *The Last Word.*

National Academy of Sciences dictates what science is in the face of no evidence:

> "Science is a way of knowing about the natural world. It is limited to explaining the natural world **through natural causes**" [my emphasis]
> National Academy of Sciences.
> Http://www.nap.edu/books/0309063647/html/

This book and others like it should never have to be written. The idea that all of the complex systems we see around us are the result of random, serendipitous chance happenings is so absurd that it makes us laugh. However, after decades of propaganda, people generally accept evolution as a fact. I wrote this book so that I could organize my thoughts and gain a clear understanding of the issues surrounding this politically charged and religiously motivated subject of evolution. I had to ask myself: why do I cringe every time I hear the narrator say those awful words, "millions of years ago"? It annoys me to distraction. The words are phrased as if they were a statement of fact. What has plagued me about the subject is: if creation is so reprehensible and worthy of ridicule, why are the evolutionists so afraid of letting schools make the comparison between creation and evolution and let "the chips fall as they may"? The answer is obvious.

How did the first cell divide without the extremely complex information of DNA, and what is the force that makes it operate? What programmed intelligence tells it how and when to divide? Where did the cell's nutrition originate? Where are all the transitional fossils that would prove evolution? Virtually all of the supposed transitions have turned out to be fabrications or outright lies (see Appendix B

- Unsubstantiated Claims of The Evolutionists)? We live in a fine-tuned universe; how did it become so finely tuned? There are many other questions that need answers. It is my intent, through research, to give plausible answers to these and others. Remember, evolution is a religion, just like Islam, Hinduism, and other religions. Attack their god, and you will suffer the consequences.

One thing is very clear: if an idea is postulated long enough, it will become fact where no fact exists. Hitler's propagandist, Joseph Goebbels, knew this full well and used it extensively to fool the German population. In fact, it is what politicians do when they get caught. They lie and deny until the prosecution stops.

> **Science now knows that many of the pillars of Darwinian theory are either false Or misleading. Yet biology texts continue to present them as factual evidence of Evolution. What does this imply about their scientific standards?**
>
> --Jonathan wells

Example: the separation of church and state. Nowhere in the US Constitution or Bill of Rights do those words appear, nor is there any concept of that idea. If you ask the average person on the street, they will tell you it is in the US Constitution. The idea of man-made global warming is spoken of as if it were a fact. Thousands of climate scientists and astrophysicists disagree. Harry Reid, Senate Democrat leader at the time, said, "Global warming is not debatable." Is he a scientist? For that matter, is Al Gore a climate scientist? We can see from history that there have been times when the earth was warmer. In fact, the result was a great increase in agricultural success. Now, the push is on; Obama and the Pope call it one of the most important crises of our time. But the truth is, it is strictly a political move and not a scientific one. The subject of evolution is the same. The idea of evolution is spoken of as fact, and anyone who disagrees is instantly branded as religious and dismissed as ignorant, especially if you are a believer in the Almighty God.

There are those who have a great deal of faith in Darwin's idea. I call it an idea because there is not enough scientific evidence for it to even be called a theory, let alone a hypothesis. That statement will surely make some Darwinists very angry. They have a lot invested in their belief. And yet, many formerly devout evolutionists, when they questioned the status quo, have turned to creation as the only plausible explanation for the origin of everything we see. The evolutionists are putting their faith in a belief that there is no God. If they are right and everything we see around us just happened by some explosion and billions of years of evolution, then life has no meaning. You should try to get everything you can now, live life to the fullest, take everything you want; there are no consequences after you are dead, satisfy all your desires now. Taken to its logical conclusion, survival of the fittest takes on a whole new meaning. As a matter of fact, the two boys who did the shooting at Columbine High School wore T-shirts that read, "Natural Selection" (from the video game). Adolf Hitler used the concept to justify killing the Jews and those not physically or mentally up to his standards. Now, the socialists of our current day are trying to move us in a similar direction. The interesting thing is that extremist Muslims believe that anyone who does not believe in Allah is an infidel and must convert, die, or pay an enormous tax. They center most of their killing right now on Muslims who are not radical enough, in addition to Christians and Jews.

I am not one to follow the masses into a burning building just because the crowd said it's warm in there. Belief in evolution is like that... trust us, we are the prevailing thought, we are "the scientists." Anyone who disagrees is immediately ostracized. The attack is not on the fact of Creation or Intelligent Design but against the person with the dissenting belief. This was disturbing to me. It reminded me of the tactics used by the Communists in their propaganda machine: discredit the person, destroy their reputation and career, but never debate the issues. (If you doubt what I say, watch the movie *Expelled: No Intelligence Allowed* by Ben Stein.)

We must remember, however, that well-meaning scientists are directed by "the elite" to look for naturalistic solutions to the origins of life only. That explains some of the totally ridiculous theories they put forth as science. They are not looking for what is true; truth does not matter; they must provide naturalistic solutions or suffer the humiliation and expulsion from the "scientific elite community." Scott Todd said,

> ***'Even if all the data point to an intelligent designer, such a hypothesis is excluded from science because it is not naturalistic"*** Scott Todd, correspondence to Nature, 30 Sept. 1999

How did this all begin? Who made that the rule of the day? ...A religion where faith in random chance and undirected natural development explains everything, from the origin of life to the cosmos. I think you will agree, if you are honest and not biased by the propaganda, that what we see around us cannot be explained by chance, but is designed by the Creator. Randomness and chance do not create order!

I make no excuses for my belief that God created the world in a literal six days or that this book is written to provide the facts that there is a God of Creation. I think there is sufficient evidence for that belief. I recommend the book "*In Six Days - Why Fifty Scientists Choose To Believe In Creation*" - Master Books Publishing - as proof. If your god had to use some form of evolution like the theistic evolutionist believes, then he is not much of a God. I serve a God who is almighty!

If I am wrong and the evolutionists are correct, when we die, everything ends, and I have wasted my time; but if they are wrong, they could spend infinity in darkness, separated from God, in a place where time does not exist. I do not pretend to know what hell is like, but I was once left in a completely dark place with no sound, totally alone and without hope. That experience had a very profound effect. I realized for a brief moment what it may be like to be separated from God in outer darkness with no hope of ever escaping, and I was not suffering the flames and torment of hell. I pray no one ever goes there. If the evolutionists are wrong, they have no idea what is in store for them. Infinity cannot be comprehended by our limited experience. Once we die, we immediately enter a new dimension, which we cannot explain or understand with our limited minds, living only in four dimensions: height, width, depth, and time.

"An open letter concerning Religion and Science"

I came across an article that stated 12,000 pastors had signed a document which claimed that the Bible is fundamentally a book about how to get along with others but has no scientific relevance. I was outraged and disappointed that there would be that many so-called men of God who would deny their Creator. If you study the early scientists (Newton, Kepler, and many others), you will find that the Bible had a profound effect on their discoveries. We now know that Columbus, directed by the Holy Spirit and the book of Isaiah, was convinced he could travel west and reach land. We think his belief was based on Isaiah 40:22. In this scripture passage, the prophet Isaiah made reference to "the circle of the earth." Here is what he said in his diary:

> "It was the Lord who put it into my mind,
> (I could feel His hand upon me), the fact that it would be possible to sail from here to the Indies. All who heard of my project rejected it with laughter, ridiculing me. There is no question that the inspiration was from the Holy Spirit, because He comforted me with rays of marvelous inspiration from the Holy Scriptures.......
>
> I am a most noteworthy sinner, but I have cried out to the Lord for grace and mercy, and they have covered me completely. I have found the sweetest consolation since I made it my whole purpose to enjoy His marvelous Presence.
>
> For the execution of the voyage to the Indies, I did not make use of intelligence, mathematics or maps.
> It is simply the fulfillment of what Isaiah* had prophesied......
>
> No one should fear to undertake any task in the name of our Savior, if it is just and if the intention is purely for His holy service. The working out of all things

> has been assigned to each person by our Lord, but it all happens according to His sovereign will, even though He gives advice. Oh, what a gracious Lord, who desires that people should perform for Him those things for which He holds Himself responsible! Day and night, moment by moment, everyone should express their most devoted gratitude to Him."
>
> - From *"Columbus's Book of Prophecies"*.
>
> This book is only available in Spanish. It has never been published in the English-speaking world. Some excerpts have been privately translated into English by August J. Kling. He quoted the passage above from Columbus's diary in an article which appeared in the October 1971 issue of "The Presbyterian Layman"

I don't know anyone who signed the open letter concerning religion and science, but they could not have had a personal relationship with God and his Word (Jesus Christ) and make that claim. The following is the letter they signed:

> Within the community of Christian believers, there are areas of dispute and disagreement, including the proper way to interpret Holy Scripture. While virtually all Christians take the Bible seriously and hold it to be authoritative in matters of faith and practice, the overwhelming majority do not read the Bible literally, as they would a science textbook. Many of the beloved stories found in the Bible - Creation, Adam and Eve, Noah and the ark - convey timeless truths about God, human beings, and the proper relationship between Creator and creation expressed in the form capable of transforming these truths from generation to generation. Religious truth is of a different order from scientific truth. Its purpose is not to convey scientific information but to transform hearts.

The secular humanists claim that those who are "religious" were or are flat Earth believers. This is far from the truth because that thinking was the result of a book by Washington Irving. The true believers never believed that, but it is their way of putting Christians in an unfavorable light, making them appear stupid.

According to recent research, the modern view that people of the Middle Ages believed that the Earth was flat is said to have entered the popular imagination in the 19th century, thanks largely to the publication of Washington Irving's fantasy The Life and Voyages of Christopher Columbus in 1828.

Russell, Jeffrey B.. "The Myth of the Flat Earth". American Scientific Affiliation.
Http://www.asa3.org/ASA/topics/history/1997Russell.html.
Retrieved 2007-03-14.

After seeing that 12,000 pastors signed that letter announcing that God did not create the world, I decided that the issue must be addressed. If God is not the Creator, then why believe at all? Is Christianity nothing but a belief in stories and fanciful guidelines? The whole basis of our belief in Christ's death and resurrection is meaningless. This then is the beginning stages of godless humanism and the ultimate objective of our greatest nemesis: the Son of Perdition, Lucifer. This book is intended to build your faith, to present the facts that God is the Creator of the universe, and to change the minds of the 12,000 or, at least, those attending their churches. It is not my intent to convince the atheistic evolutionist because their bias and hatred of all things Christian is virtually impenetrable, and only God can reach them.

There is an all-out attack on anyone who would question the theory. The attack is always predicated on the fact that our beliefs are based on a belief in God; it is therefore religiously based and not scientific. This is far from the truth, which I will illustrate with many sources. The evolutionists are just as religious as we are, but their God is undirected chance. I hope to prove that it takes far more faith to believe in chance in the light of little or no evidence than in a Creator God.

Creation science is not given a proper place in our classrooms and universities. I am not suggesting that evolution not be taught; far from it, I would love to see the comparison taught so that the student can make up their own mind. However, the ACLU and people like Eugenie Scott have made sure that ONLY evolution is taught and that there is no mention of Intelligent Design, Creation, or discussion of the many flaws in evolution. They base this on the premise that Creation and ID are religious, and they invoke the "separation of church and state" argument. Just like Darwin's ideas, they are baseless.

Bill of Rights - *Amendment I - Congress shall make no law respecting an establishment of religion, or prohibiting the free exercise thereof; or abridging the freedom of speech, or of the press; or the right of the people peaceably to assemble and to petition the Government for a redress of grievances.*

Teaching an opposing view to evolution does not require that Congress establish "a religion." However, Congress, by continuing to allow one religion to rule in our educational system and denying the other, is, for all practical purposes, unintentionally establishing a state religion. What is that religion that has free rein in our educational system? Humanism. Humanism is as much a religion as Christianity. I do not know how Congress can allow the state to make these beliefs preeminent and deny all others. It is obvious that their agenda is to make the United States a secular nation. They have succeeded in establishing the religion of secular Humanism in our public schools. Now there is a daily attack on Christianity in all areas of public life, from the pledge of allegiance to denying entrance to the University of California by graduates of Christian schools.

The evolutionist dare not allow evolution to be compared with Intelligent Design or Creation Science (I make the distinction between Intelligent Design and Creation because Creation requires a belief in God the Creator, while Intelligent Design only requires some outside intelligence). When you begin to compare evolution to Creation, the evolutionary theory falls short immediately. There are numerous cases where there is outright fraud in attempting to prove the theory. This book will discuss not only the shortcomings and fraud of Darwin's followers, but also why Creation can be the only true answer.

Tim Cocklin

The coming delusion

The evidence for design is so overwhelming that the evolutionist, if he or she is intellectually honest, must concede that everything around us has its origin with a designer.

The more science discovers, the more we see design. As a matter of fact, we would not even be having this conversation if it were not for atheism and secular humanism. These two ideologies drove scientists to attempt to find naturalistic solutions for the folly of evolution. This caused evolutionists to come up with ridiculous hypotheses to prove their point. In the end, these ideas are either proven to be a complete fabrication, outright lies, or ill-conceived ideas based on a faulty premise.

There are many things we do not understand. Scientists now believe that we live in ten dimensions, four of which are knowable: height, depth, width, and time, and six or more which are not knowable. This is referred to as string theory if you want to research it on your own.

After the resurrection of Jesus, He was able to defy the laws of physics by moving through solid walls, and yet they could feel the nail prints in His hands. He would be there one moment and not the next. It is my belief that in His resurrected state, He was able to move in dimensions we cannot know at this time.

> *Then the same day at evening, being the first day of the week, when the doors were shut where the disciples were assembled for fear of the Jews, came Jesus and stood in the midst, and saith unto them, Peace be unto you. 20 And when He had so said, He shewed unto them His hands and His side. Then were the disciples glad, when they saw the Lord. John 20:19-20 (KJV)*

> *And after eight days again His disciples were within, and Thomas with them: then came Jesus, the doors being shut, and stood in the midst, and said, Peace be unto you. 27 Then saith He to Thomas, Reach hither thy finger, and behold my hands; and reach hither thy hand, and thrust it into my side: and be not faithless, but believing.*
> *John 20:26-27 (KJV)*

> "When a scientist of Crick's caliber feels he has to invoke undetectable spaceman, it is time to consider whether the field of prebiological evolution has come to a dead end."
> Philip Johnson, *Darwin on Trial,* p. 111.

The Enemy of God will do anything to be like God. Now, if UFOs are real, and far too many people have seen them to discredit their testimony, where do they originate? Although we do not know for sure, the fact that they can break the laws of physics, turning at 90-degree angles at extremely high speeds, and the fact that they can be there and then disappear gives us a clue that they operate in a dimension we cannot yet comprehend.

It is pure speculation, but I am thoroughly convinced that in the future, the overwhelming evidence of design will not continue to be ignored and ridiculed. Men like Steven J. Meyer, Jonathan Wells, William Dembski, Jay Richards, Guillermo Gonzales, and many other Intelligent Design proponents, as well as the multitude of Creation Scientists who have been viciously condemned, will be vindicated just like the many scientists before them, Galileo, for example. That is, unless the federal government outlaws open scientific research for neo-Darwinism. There are more and more scientists with substantial credentials proposing Intelligent Design. The theory of evolution must, by necessity, go the way of bloodletting and a flat Earth as an ill-conceived 19th and 20th-century concept.

What does the Evolutionist believe?

To illustrate the bias the evolutionists have, we quote what its most stringent proponents are saying.

> 'Even if all the data point to an intelligent designer, such a hypothesis is excluded from science because it is not naturalistic" Todd, S.C. correspondence to Nature 401(6752):423, 30 Sept. 1999;

The evolutionist wants freedom from accountability to God, this is the reason they fight so hard to keep evolution alive otherwise they lose the ability to do whatever they want. Listen to Aldos Huxley, grandson of Darwin's bulldog, Thomas Huxley.

> "I had motives for not wanting the world to have meaning; consequently assumed it had none, and was able without any difficulty to find satisfying reasons for this assumption. The philosopher who finds no meaning in the world is not concerned exclusively with a problem In pure metaphysics; He is also concerned to prove there is no valid reason why He personally should not do as He wants to do . . For myself, as no doubt foremost of my contemporaries, the philosophy of meaninglessness was essentially an

> instrument of liberation. **The liberation we desired was simultaneously liberation from a certain political and economic system and liberation from a certain system of morality objected to the morality because it interfered with our sexual freedom."** Aldous Huxley."*Confessions of a Professed Atheist*," Report: Perspective on the News, Vol. 3, June 1966, p. 19 [my emphasis]

We are all affected by our worldview. The Creationist has a view that the God of the Bible created everything. The neo-Darwinist has the worldview that there is no God and that everything happened by random fortuitous chance events and natural selection over eons of time. The whole premise of neo-Darwinism is based on the pretentious belief that all creation can be determined by natural causes. The question is, who decided that only natural causes would be the definition of true science?

To believe in Intelligent Design takes virtually no faith. Look at the evidence; it is easy to see the hand of an intelligent designer. The symbiosis of living things and the finely tuned universe vociferate design. You have to be deaf and blind not to comprehend it. However, note that this belief does not require you to believe in the God of the Bible. (*See the origins comparison chart in Appendix A.*) To believe that the God of the Bible is the designer does require faith, since He can only be known to those who have had a genuine encounter with Him in new birth (1 Corinthians 2:14). On the other hand, the Darwinist must have blind faith to believe that everything we see just happened in a series of serendipitous chance happenings in small progressive steps, especially since there is very little evidence for it. Statistically, it is impossible for these things to have come about by mere chance; it breaks the law of biogenesis § and many other well-established laws. You will discover that the probability of life coming from non-life is zero.

The Big Bang does not explain it; chance mutations over eons of time do not explain it, and natural selection does not create new systems; that is macro-evolution (a dog is still a dog kind even if it came from a wolf). By their admission, the neo-Darwinist must find ways to make natural causes the answer for everything we see, from animals to plants to the cosmos. Stephen J. Gould, one of the strongest advocates of evolution, admitted that we don't see changes from one species to a new one, only slight variations. We agree this is known as micro-evolution. His advice to his fellow evolutionists is that if it is not evolution, then do not talk about it.

> "Every paleontologist knows that most species don't change. That's bothersome ... brings terrible distress. ... They may get a little bigger or bumpier. But they remain the same species and that's not due to imperfection and gaps but stasis. And yet this remarkable stasis has been ignored as no data. If they don't change, it's not evolution so you don't talk about it." Gould, Stephen J. in Lecture at Hobart and William Smith College. February 14, 1980

Gould was right to tell them not to talk about things that do not support their position. What a believer in evolution typically does is change the meanings of terms in the same conversation. For instance, the word "change" to the evolutionist broadly means entities are evolving from one creature or adding a feature to another through small steps and natural selection. This has never been legitimately proven. There have been numerous fraudulent attempts to prove it, but in the end, they are deemed false. However, the same term "change" will be used to justify their point of view when small changes occur in an entity. Look at the change over time of the finches in the Galapagos Islands. When the weather does not produce enough rain, the number of finches with long beaks will increase. However, when the weather changes again, so do the number of short beaks. They claim this is proof of evolution. How is that evolution? You started with finches, and you ended with finches. The short and long beak finches did cross-breed, creating other beaks, and yet they are still all finches. They didn't even become some other type of bird.

Summary: What does the Darwinist believe?

(See appendix D for Creation Science Critique of Neo-Darwinism)

1. There is no Creator or Intelligent Designer all things came about through ***natural UNDIRECTED causes.***

2. A "Big Bang" event. There currently many evolving theories and these are constantly changing, because of new discoveries which prove their theory incorrect.

3. Chance fortuitous mutations that result in small incremental improvements that are retained by natural selection. Although, now because of the evidence; they have proposed that punctuated equilibrium is the answer.

4. Long periods of time (millions and billions)

I. There is no Creator or Intelligent Designer

Gould was right to tell them not to talk about things that do not support their position. What a believer in evolution typically does is change the meanings of terms in the same conversation. For instance, the word "change" to the evolutionist broadly means that entities are evolving from one creature or adding a feature to another through small steps and natural selection. This has never been legitimately proven. There have been numerous fraudulent attempts to prove it, but in the end, they are deemed false. However, the same term "change" will be used to justify their point of view when small changes occur in an entity. Look at the change over time of the finches in the Galapagos Islands. When the weather does not produce enough rain, the number of finches with long beaks will increase. However, when the weather changes again, so will the number of short-beaked finches. They claim this is proof of evolution. How is that evolution? You started with finches, and you ended with finches. The short- and long-beaked finches did cross-breed, creating other beak types, and yet they are still all finches. They didn't even become some other type of bird.1 Corinthians 2:14 says:

> *"But the natural man receiveth not the things of the Spirit of God: for they are foolishness unto him: neither can He know [them], because they are spiritually discerned".*

Because God can only be understood spiritually, the spiritually dead are not capable of knowing God intimately. The agnostic will say "God cannot be known", so we cannot attribute the creation of life to Him. The atheist denies that there are any gods at all. Professor William Provine, Cornell University, said,

> "… There are no gods, no purposes, no goal-directed forces of any kind. There is no life after death. When I die, I am absolutely certain that I am going to be dead. That's the end for me. There is no ultimate foundation for ethics, no ultimate meaning to life, and no free will for humans." Provine, W.B. Origins Research 16(1):9, 1994.

If they believe there is no free will and they promote it fully, then the next generation will be the most violent and self-centered in human history. We already see the fruits of that belief, with the mass killings of people, the Black Lives Matter movement, and the recent killings of police. There is a total disregard for law and order. Not having free will means that you are only directed by chemical reactions and are, therefore, not responsible for your actions. "My chemistry made me do it" or "my environment caused me to be this way." If that does not terrify you, you do not understand the ramifications of that belief. I just read an article about a man (52). He saw two girls fighting in the back of his truck. He came out of his house and asked them to leave. At that point, the crowd of teenagers turned on the man and beat him senselessly. He had broken his eye sockets, nose, and ribs. He was in critical condition. Is this an indication of how years of teaching that there is no free will leads to no absolutes? In McKinney, Texas, teenagers were asked to leave by a community security guard because they were not legally (as defined by the homeowners association) on the property. When they refused, he called the police. When the police demanded that they all stop and sit down, most refused, and all hell broke loose. There is increasing lawlessness, and we will see more of it in the coming days. I could give numerous examples, but that is for another publication.

> *And then many will be offended and repelled and will begin to distrust and desert [Him Whom they ought to trust and obey] and will stumble and fall away and betray one another and pursue one another with hatred. Matthew 24:10 - Amplified Version - Zondervan*

Wernher Von Braun, who has been called the father of the American space program, was not convinced of the idea of evolution, He declared that design is clearly evident everywhere and that there is order and a plan.:

> "For me, the idea of a creation is not conceivable without invoking the necessity of design. One cannot be exposed to the law and order of the universe without concluding that there must be design and purpose behind it all. In the world around us, we can behold the obvious manifestations of an ordered, structured plan or design . . .
>
> "Some people say that science has been unable to prove the existence of a Designer. They admit that many of the miracles in the world around us are hard to understand, and they do not deny that the universe, as modern science sees it, is indeed a far more wondrous thing than the creation medieval man could perceive. But they still maintain that since science has provided us with so many answers the day will soon arrive when we will be able to understand even the fundamental laws of nature without a Divine intent. They challenge science to prove the existence of God. But must we really light a candle to see the sun? . . ." Ibid

The idea of a Creator is not new or just the fabrication of religious people. Lao Tzu, the founder of Taoism, wrote in 600 B.C.

> "Before time, and throughout time, there has been a self-existing being, eternal, infinite, complete, omnipresent. Outside this being, before the beginning, there was nothing." Lao-tzu, Tao-te-ching, circa 600 B.C. (tr. Leon Wieger. English version by Derek Bryce. 1991. Llanerch Publishers,

The problem for the evolutionist is that they must continue to change their story as new information is discovered. Stephen Hawking said that the universe is without boundary and always existed and, therefore, there is no need for a Creator. This would break the Laws of Thermodynamics.

> "The idea that space and time may form a closed surface without boundary also has profound implications for the role of God in the affairs of the universe? If the universe is completely self-contained, having no boundary or edge, it would have neither beginning nor end: it would simply be. What place, then, for a Creator?

Unfortunately for Mr. Hawking, new information has concluded that the universe is finite. An article in National Geographic News shows new data and finds the universe is finite.

> Now, a new study of astronomical data only recently available hints at a possible answer: The universe is finite and bears a rough resemblance to a soccer ball or, more accurately, a dodecahedron, a 12-sided volume bounded by pentagons. If proven by further evidence and scrutiny, the model would represent a major discovery about the nature of the cosmos. - Sean Markey - National Geographic News - October 8, 2003

Although, at this time, the new data is being debated it shows that man's theories are subject to change with new discoveries. I predict that the more science discovers, the more the Creationist will be proven correct. Will they get the credit? Not likely!

2. The Big Bang

The cosmos is not the focus of this series of books. Suffice it to say that one of the laws of physics is: matter can neither be created nor destroyed. The evolutionist will tell you: there was nothing, and then it exploded. There are many variations of this, but if questioned long enough, the evolutionist will have to admit that it makes no sense. I believe in a Big Bang, but it came from the spoken words of the Almighty. After years of propagating the story of the "Big Bang," many of its proponents are abandoning the idea. However, new theories are being created every day to prove their point.

3. Chance - fortuitous, undirected, positive mutations

Before we continue, we need to set something straight. When we speak of evolution, we mean the belief that all organisms come from a single ancestor. That, over eons of time, by small incremental changes, we can account for all living things. We freely admit that there are variations in plants and animals; that is built into the DNA, called micro-evolution, and controlled by genetic boundaries. For instance, you have all the necessary genetics to have skin that is black, white, or anything in between. However, never in nature do we see an apple become a bean or a cow become a whale; that is called macro-evolution, and it has never been proven.

> "Many evolutionary propagandists are guilty of the deceitful practice of equivocation, that is, switching the meaning of a single word (evolution) part-way through an argument. A common tactic, 'bait-and-

> switch,' is simply to produce examples of change over time, call this 'evolution,' then imply that the GTE [General Theory of Evolution] is thereby proven or even essential, and creation disproved. The PBS Evolution series and the Scientific American article are full of examples of this fallacy". - Jonathan Sarfati, Ph.D. F.M. Refuting Evolution 2, Chapter 1, Argument: Creationism is religion, not science

The theory of evolution has a major problem when dealing with chance fortuitous mutations. First, it has never been proven that mutations lead to improvements in an organism. We will discuss numerous mutations in this book; none of them are helpful. Second, mutations never add new functional information, and third, if a mutation is to be passed on to the next generation, it must occur in the reproductive cells. How would that happen?

Mutations do not lead to improvements: When I first started as a programmer at Computer Language Research, I modified existing PL/1 programs to make improvements or modifications based on new tax law. Be assured that unless you understand what is going on in the current iteration, you are likely to make things worse. Once the program has run, you will be able to tell if the modifications work. If they do not produce the desired result, you simply go back and modify the program until the desired result occurs. Blind chance, on the other hand, making mutations to your DNA would have no way of knowing how the mutation will affect the whole system. Working in small incremental steps would most certainly destroy the entity before it gets started.

I.L. Cohen makes a good point: there is a very fine balance in the DNA. Upsetting this balance does not bring about positive effects.

> "Summing it up, and contrary to the oft-repeated evolutionist claims, a random mutation is not an enhancing factor -- it is almost invariably a destructive one. It interferes with the exquisitely

> fine balance and purposefulness of the hundreds of millions of nucleotides within a given program. To propose and argue that mutations even in tandem with 'natural selection' are the root-causes for 6,000,000 viable, enormously complex species, is to mock logic, deny the weight of evidence, and reject the fundamentals of mathematical probability." - Cohen, I.L. In *Darwin Was Wrong - A Study in Probabilities.* New Research Publications, Inc. New York, NY (1984), p.81.

Biologist Ranganathan agrees that changes in a highly ordered system do not make a system better.

> "Genuine mutations are very rare in nature. Most mutations are harmful since they are random, rather than orderly changes in the structure of genes; any random change in a highly ordered system will be for the worse, not for the better. For example, if an earthquake were to shake a highly ordered structure such as a building, there would be a random change in the framework of the building, which, in all probability, would not be an improvement.". G. Ranganathan, *Origins?*, Pennsylvania: The Banner Of Truth Trust, 1988

The real question we need to ask the evolutionist is: ***how*** did things evolve if not through chance mutations? If we follow the scientific method, we should be able to see undeniable proof that the hypothesis is true before giving it a title like 'theory.'

It is clear from the evidence that most mutations cause sickness and deformity in the phenotype (the outward expression of an organism) and not new forms of life. Ernst Mayr makes this case in *The Multiplication of Species* (Belknap Press - Harvard University Press, Cambridge, MA (1970), 2nd edition)

While mutations can indeed produce extreme genetic abnormalities, these changes typically result in organisms so malformed that they are essentially "hopeless" from an evolutionary standpoint. Such drastic mutations often disrupt the organism's balance and function to the point that they have no realistic chance of surviving natural selection. In fact, the more significant the mutation's impact on an organism's traits, the more likely it is to harm rather than help. Believing that a major mutation could give rise to a viable new species capable of thriving in a different ecological niche is comparable to believing in miracles. Furthermore, the idea that such an organism could find a compatible mate and become reproductively isolated from the original population presents challenges that are, in Mayr's view, practically impossible to overcome.

The evolutionists base their belief on the idea that there have been successive beneficial mutations over long periods to explain the transition from one form to another until we arrive at man. But when asked to explain the lack of evidence and even the current observable facts, the evolutionist is deafeningly silent. David Demick points this out in his book, *The Blind Gunman.* In recent years, scientists have documented thousands of human diseases linked to genetic mutations, and the list continues to grow. One medical genetics reference alone includes around 4,500 distinct genetic disorders. Some conditions that were once thought to be single syndromes—like Marfan's syndrome—are now known to be genetically diverse, meaning they result from various different mutations. Given the overwhelming number of harmful mutations known to cause disease, one might expect that if macro-evolution were true, there would also be clear examples of beneficial mutations. Such positive mutations would not only be essential for driving increased biological complexity but would also need to counterbalance the many harmful ones. Yet, despite the abundance of negative examples, evolutionary scientists have offered very few well-documented cases of beneficial mutations, leading to noticeable silence on this crucial point.

If the evolutionist is honest and follows the scientific method, then the death knell to the hypothesis would be the fruit fly experiments, which have ***never*** given rise to a new species even after more than 50 years. **Gordon Taylor said:**

> "In all the thousands of fly-breeding experiments carried out all over the world for more than fifty years, a distinct new species has never been seen to emerge." Taylor, Gordon in *The Great Evolution Mystery.* Harper & Row, New York, NY (1983), p.34.

The scientific method is as follows: 1. Observation 2. Hypothesis 3. Prediction 4. Testing and Results. It is clear that a fruit fly is a fruit fly, no matter what mutations you inject. I guess they have never heard the old saying, "doing something over and over, expecting different results is insanity." It is not the focus of this series, but there is an agenda underway to change the genetics of animals and humans known as "trans-humanism." It is worth your time to check it out, although there are a number of irrational theories.

Pierre-Paul Grassé was a supporter of the French tradition of Lamarckism. Although Lamarckism has long since been abandoned by everyone because it is so easily proven false. Jean-Baptiste Lamarck proposed that individual efforts during the lifetime of the organisms, such as giraffes stretching to reach higher branches for food, would cause their necks to grow, and this new longer neck would then be passed to the next generation. It would be the same as saying that because I am a bodybuilder and I have reached a certain level of strength, my offspring will have the same. We know DNA does not carry those attributes in the genes. However, Grassé did come to the realization that mutations are not cumulative in generations; they only modify what preexists, and if they do occur, sickness and death soon follow.

> "Mutations, in time, occur incoherently. They are not complementary to one another, nor are they cumulative in successive generations toward a given

> direction. They modify what preexists, but they do so in disorder, no matter how.... As soon as some disorder, even slight, appears in an organized being, sickness, then death follow. There is no possible compromise between the phenomenon of life and anarchy." - Pierre-Paul Grassé, *Evolution of Living Organisms*, Academic Press, New York, 1977, p. 97, 98.

This is consistent with what we observe in mutations. For example, in the case of an inherited disease due to the faulty metabolism of phenylalanine is necessary for growth in infants and nitrogen equilibrium in adults [How would undirected chance know this and account for it?]. Normally, the production of the protein has 451 amino acids in its chain, with arginine at the 408th position. If that 408th position is miscoded, you get phenylketonuria (PKU) or the potential for mental retardation. You can see from the illustration below that if just one base pair is coded incorrectly, you have a potential disaster! This illustrates how very precise God is in His creation, or most of us would be mentally retarded. To credit this stringent, finely tuned design to random chance is an insult to our intelligence. We will dig deeper into the function of DNA and the cell later.

	Normal	PKU
DNA at codon* 408	x x x **C** G G x x x	x x x **T** G G x x x
	x x x **G** C C x x x	x x x **A** C C x x x
Amino Acid at 408	arginine	tryptophan

*Note: a codon is a three-letter nucleotide in DNA that codes for a specific amino acid. (See the chapter on DNA for more information)

Personally, I feel silly even writing this book. Evolution is so full of bias and prejudice that I cannot believe we still talk about it. For the evolutionist to imply that evolution is science and creation science is not is absurd. There is a growing number of evolutionary scientists who cannot accept Darwin's ideas as valid anymore; there is too much evidence to the contrary.

Let's look briefly at the famous Stanley Miller experiment. The objective was to prove that amino acids, the building blocks of protein, could arise by purely naturalistic means. Much positive publicity has been given to the experiment, and it is still used in textbooks as proof for evolution, but does it prove chemical evolution? Miller used a combination of gases that he assumed had existed in the early atmosphere, such as ammonia, methane, hydrogen, and water vapor (but no oxygen). He heated the temperature in his "environment" to 212°F (100°C) for a week while also providing an electrical current to act like lightning. At the end of the week, Miller measured the chemicals in the mixture at the bottom of the jar (his cold trap) and observed that he had synthesized three of thc 20 amino acids constituting the building blocks of proteins. However, later the experiment was invalidated for many reasons:

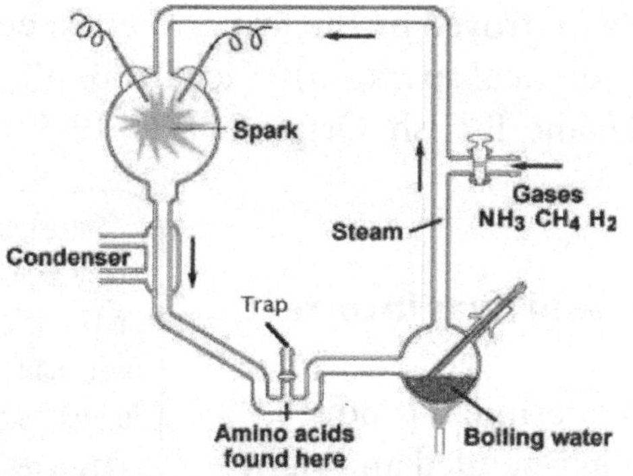

PUBLIC DOMAIN INDIANA.EDU

'Ihe wrong atmosphere was used: In 1982, scientists agreed that instead of methane and ammonia in the primitive atmosphere, there must have been nitrogen and carbon dioxide. Indeed, after a long silence, Miller himself admitted that the primitive atmosphere model he'd used was not realistic.

The American scientists J.P. Ferris and C.T. Chen repeated Miller's experiment, using a mixture of carbon dioxide, hydrogen, nitrogen, and water vapor, but failed to obtain even a single amino acid molecule. Photochemistry of Methane, Nitrogen, and Water Mixture as a Model for the Atmosphere of the Primitive Earth," Journal of American Chemical Society, Vol. 97:11, 1975, p. 2964.

Invalid testing methods were used:

He used a trick to capture the results of His experiment in a trap. As the chemist Richard Bliss has stated, "Without this cold trap, the chemical products would be destroyed by the [experiment's] energy source (electrical sparking)." Richard B. Bliss, Gary E. Parker, Duane T. Gish, Origin of Life, 1979, pp. 14-15

"Since molecular biologists began to appreciate the sequence specificity of proteins and nucleic acids, many calculations have been made to determine the probability of formulating functional proteins and nucleic acids at random. ... Such calculations have invariably shown that the probability of obtaining functionally sequenced biomacromolecules at random is, in [Nobelist Ilya] Prigogine's words, 'vanishingly small'. ... Chance is not an adequate explanation for the origin of biological complexity and specificity." - Stephen C. Meyer, The Message in the Microcosm: DNA and the Death of Materialism,"

The result was incorrect:

The experiment produced both left-handed and right-handed amino acids. The appearance of right-handed amino acids would negate any potential for proteins. All proteins in a living entity are left-handed.

All proteins in a living entity are left-handed.

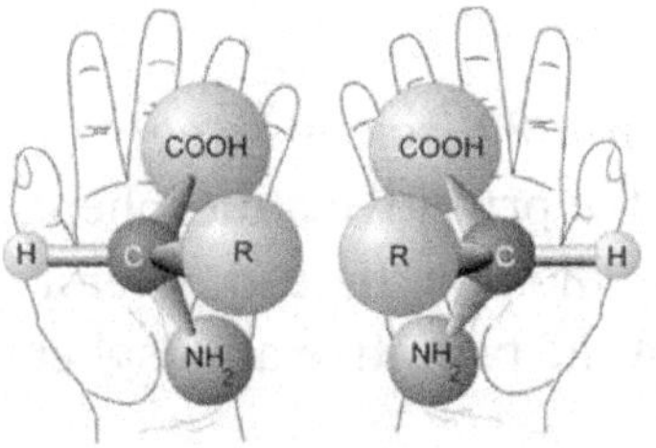

The experiment was a complete failure and yet it is still taught in schools and universities that Miller proved the case for evolution. What Miller did produce was a chemical "goo" made up of 85% tar, 13% Carboxylic acids (toxic) and 2% amino acids. When amino acids are synthesized in the laboratory, they always get 50% L and 50% right-handed amino acids.

Left handed (Levo) and right-handed amino acids (Dextro) are a mirror image of each other, just like your right hand is a mirror image of your left. However, if just one right-handed amino acid is found in an amino acid chain, the chain is worthless because all proteins must fold in a very specific way to be viable. Of the twenty different amino acids in the human body, only one is not left or right: glycine. The other nineteen can be either right or left. (See Appendix B vocabulary for a list of the amino acids). All amino acids in proteins are 'left-handed', while all sugars in DNA and RNA, and in the metabolic pathways, are 'right-handed'. *Origin of life: the chirality problem* by Jonathan Sarfati - creationwiki.org

Each type of amino acid, when found in nonliving material or when synthesized in the laboratory, comes in two forms that are chemically equivalent. About half of these amino acids can be described as 'right-handed,' and the other half as 'left-handed.' Each structure is a mirror image of the other. However, the amino acids that comprise the proteins found in living things, including plants, animals, bacteria, molds, and even viruses, are essentially all left-handed. No known natural process can isolate either the left- or the right-handed variety. The mathematical probability that chance processes could produce just one tiny protein molecule with only left-handed amino acids is zero.

> "A similar observation can be made concerning sugars which, in living organisms, are essentially all right-handed . . Based on our present understanding, natural processes produce equal proportions of left- and right-handed sugars. Since the sugars in living things are almost all right-handed, our present understanding leads to the conclusion that natural processes probably did not produce life."—Walter T. Brown, *In the Beginning (1989), p. 8.*

Natural Selection

Natural selection explains nothing about the causes of changes within an organism's genes. How would natural selection account for the overwhelming variety in nature, such as the color of a bird's egg? I remember growing up on a farm where we had many opportunities to see nature up close. We could investigate a variety of birds and their eggs, noting the different sizes and colors. How would the color of an egg make an organism more likely to survive, especially since they are attractive to the eye?

We concede that variation takes place in the various kinds (known as micro-evolution). However, we have never seen vegetables turn into fruit or cows become whales.

> "No one has ever produced a species by the mechanisms of natural selection. No one has ever got near it, and most of the current argument in neo-Darwinism is about this question." Colin Patterson, "Cladistics" Interview by Brian Leek, interviewer Peter Franz, March 4, 1982, BBC.

4. Long periods of time

> Nicolas Steno (1638-1686) is considered the father of geology and stratigraphy. In 1666, two fishermen caught a shark and it was brought to Steno to study. When He examined the teeth He noticed that they appeared to be identical to stony objects found embedded in rock. These were known as "tongue stones". His question was, how could any solid object be encased in another solid object? In the end, He came to the conclusion that the fossil record was a chronology of different living creatures in different eras. Nicolas Steno also reasoned that rock strata were formed when particles in a fluid such as water fell to the bottom. - Wikipedia

This sounds remarkably like what would have happened in the Bible's account of Noah's flood.

William Smith (1769–1839) was an English geologist, credited with creating the first nationwide geological map. He is known as the

"Father of English Geology," although recognition was very slow in coming. At the time his map was first published, he was overlooked by the scientific community.

To further confuse the issue, Charles Lyell (1797–1875) published his theory of radical uniformitarianism. Uniformitarianism, in the philosophy of science, assumes that the same natural processes that operate in the universe now have always operated in the universe in the past, and at the same rates; and that the same laws of physics apply everywhere in the universe. Its methodology is frequently summarized as "the present is the key to the past," because it holds that all things continue as they were from the beginning of the world. We are now aware of the fact that from private letters to his sister, Lyell had intended to find a way to discredit Genesis and the Flood. It happens that Darwin had a copy of Lyell's work when he took his famous trip on the Beagle. The steady accumulation of minute changes over long spans of time had a very profound effect on Darwin when formulating his thoughts on evolution.

Now, here we are in the 21st century, and we are still taught that the geologic record proves evolution, when in fact, there is plenty of evidence to the contrary.

The scientific method

The scientific method is an accepted method of taking an observation to a hypothesis; then, if the predictions and tests prove true, it becomes a theory. If that theory is observed after further tests to be incontrovertible, the theory becomes a law. Let's apply one of Darwin's ideas to the scientific model. Keep in mind that all evidence is circumstantial for Creation and Evolution since we have no witness who was there to observe it. However, since the evolutionist speaks about evolution as a fact, let's see if it stands up to the scrutiny of the scientific model.

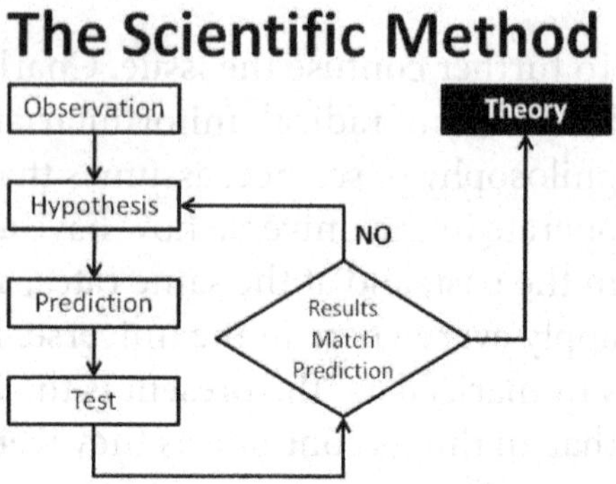

Darwin ***hypothesized*** that all creatures came from a single ancestor. He ***predicted*** that there would be millions of examples of transitions in the fossil record. His ***presupposition*** was that the accumulation of millions of years of sedimentary rock containing the fossilized remains would prove his hypothesis. The rocks would show the transition from one form to the next. Single-celled organisms will become more and more complex, from sea creatures to reptiles to birds, and mammals. Let's test this hypothesis.

To date, after more than 150 years of fossil discovery, we have yet to see any transitional forms, except the obvious frauds foisted on the scientific community as proof (see Appendix B, Unsubstantiated Claims of The Evolutionist). The one example of this fraud is IDA. Prof. Peter Brown, a paleoanthropologist at the University of New England, said:

> "The scientific study did not provide sufficient proof to consider Ida as a human ancestor, or even a primate

> ancestor. It has fingernails, something we have, as do most primates ... but they've cherry-picked particular characteristics, and they've been criticized [by other scientists] for doing that." "It's not a missing link, it's not even a terribly close relative to monkeys, apes and humans, which is the point they're trying to make," says Carnegie Museum of Natural History curator of vertebrate paleontology Chris Beard. Quotes are from Robert MorleyColumnist *A Multi-million-Dollar Monkey Fraud* June 2, 2009 | From theTrumpet.com

The prediction is false, so his hypothesis is false and must be restated or removed as a hypothesis. Nowhere in the fossil record do we find any transitional forms. Student textbooks, teachers, and professors continue to propagate this fallacy by using examples that long ago were proven to be fraudulent (see Appendix B: Unsubstantiated Claims of Evolution). We can state emphatically, then, that those who promote the scientific method do not believe it, do not understand it, or want to deliberately deceive students with fraudulent information ***to promote an agenda***. Even the famous Wikipedia gets it wrong. Here is a quote from Wikipedia on the page about transitional fossils:

> "Indeed, Archaeopteryx was discovered just two years later, in 1861, and represents a classic transitional form between dinosaurs and birds. Many more transitional fossils have been discovered since then and it is now considered that there is abundant evidence of how all the major groups of animals are related, much of it in the form of transitional fossils. Quoting Prothero, D (2008-02-27). *Evolution: What missing link?*. New Scientist. Http://www.newscientist.com/article/mg19726451.700-evolution-what-missing-link.html [my emphasis]

Sir Fred Hoyle, an evolutionist, carefully studied the two best Archaeopteryx specimens, front and back, and declared them to be faked. They discovered that the front and back slabs of each specimen, do not match. They found that an alteration had been made to the left wing, as depicted in an 1863 drawing. They concluded that the feather markings had been imprinted by hand. They also found that etching process had used cement blobs. When the scientists requested the ability to use an electronic microscope and carbon-14 dating, the museum refused and withdrew the specimens from the scientists. British Journal of Photography (March-June 1985).W.J. Broad, *"Authenticity of Bird Fossil is Challenged,"* N.Y. Times C1, C14 (May 7, 1985). T. Nield, "*Feathers Fly Over Fossil 'Fraud',*" New Scientist 1467:49-50. G. Vines, "*Strange Case of Archaeopteryx 'Fraud',*" New Scientist 1447:3.

If there were transitions, we would be overwhelmed with evidence, not just an occasional one here and there. In fact, there are no transitional fossils that have been discovered, studied, and verified. Most transitional fossils are clearly fraudulent or simply misrepresentations.

There are two schools of thought: everything happened by accident, or it was created by an infinitely more intelligent source than we could ever imagine. In this book, we are not trying to prove Creation, but to give you enough information to determine whether neo-Darwinism, Progressive Evolution, or Theistic Evolution is true. (See the origins comparison chart in Appendix

> "...Origin of the Species is under increasing attack—and not just from creationists. ...so heated is the debate that one Darwinist says there are times when he thinks about going into a field with more intellectual honesty: the used car business".
>
> Sharon Begley,
> *Newsweek*, April 8, 1985, P 80.

As we have stated before, it takes faith to believe that God created everything. There is a very sound reason we can believe it. If there was a Big Bang, it was the voice of God when He said: "Let there be light". If there is to be eons of time, it will be when God creates a new heaven and a new earth.

At the end of the day, evolution is nothing more than a philosophy. It was never observed in the past (or present) and it cannot be tested. L. Harrison Matthews said in his introduction to the 1971 publication of the Origin of Species:

> "The fact of evolution is the backbone of biology, and biology is thus in the peculiar position of being science founded on an unproven theory—is it then a science or a faith? Belief in the theory of evolution is thus exactly parallel to belief in special creation—both are concepts which believers know to be true but neither, up to the present, has been capable of proof." - L. Harrison Matthews, *Introduction to C. Darwin, The Origin of Species* (reprint, London: J.M. Dent and Sons Ltd. 1971), P. XI.

Do you believe in magic?

Just like the fairy tale of the princess who kissed the frog and the frog turned into a prince, so goes the story of evolution. You have to believe in fairy tales to believe in the whole story of evolution. According to evolution, everything we see has been a freak accident of nature, starting with the Big Bang. They have a habit of coming up with new ideas to explain the impossible, like punctuated equilibrium. For 50,000 years or so, there will be no change (equilibrium). And then, suddenly, in a very rare punctuation and by total chance, two totally different life forms will emerge: male and female. WHAT? Do they seriously expect us to believe that, especially in the light of no evidence? Sadly, people, even highly educated people, believe it.

Punctuated Equilibrium: When evolution occurs, it is localized in rare, rapid events of branching speciation (called cladogenesis). Cladogenesis is simply the process by which species split into two distinct species, rather than one species gradually transforming into another. From Wikipedia on (punctuated equilibrium).

Since Darwin's hypothesis could not account for the sudden appearance of mass numbers of fully formed species, known as the Cambrian explosion, the Darwinists had to formulate another hypothesis. To explain the sudden appearance of fully formed plants and animals in the fossil record, Gould and Eldredge came up with a new hypothesis in 1972. But where is the evidence? Are they suggesting that once every 50,000 years or so, a miracle takes place and a male and a female are born to procreate? Convenient! You can imagine just about anything, but that does not make it true or even possible. I can imagine that if I flap my arms and teach my children to teach their children to flap their arms, in a few thousand years we could grow feathers and wings and fly off. Does anyone else get the sense that they are trying to hammer a square peg into a round hole and make it fit perfectly? The question is, how and why would this occur?

For their theory to work, you would have to have multi-millions of positive mutations occurring all at once in a short span of a few thousand years, all of which are coordinated perfectly with the existing organisms. The problem is that we have never observed mutations providing positive results. And so, they hide their theory by suggesting a few thousand years between punctuations. That is used to explain why we don't see it now. But what would have caused these changes? To me, that is even more absurd than Darwin's idea of very small incremental changes. What would cause these positive mutations to occur? Once again, they rely on pure chance for these millions of fortuitous mutations, but now it happens all at once in a short burst of time.

Raymond G. Bohlin argues that advocates of punctuated equilibrium fail to provide a credible explanation for how new genetic information arises—something essential for any evolutionary theory to be complete. He notes that there is often a poor understanding of the nature and impact of mutations. In particular, mutations affecting regulatory and developmental genes are frequently discussed without acknowledging their typically harmful consequences. Unlike mutations in structural genes, which might cause minor changes, developmental mutations often have damaging or even lethal effects. Therefore, instead of resolving evolutionary challenges, punctuated equilibrium introduces additional unanswered questions. — Raymond G. Bohlin, The Natural Limits to Biological Change (Probe Ministries)

You are faced with a choice. What are you willing to believe?

Magic

Things "magically" make themselves with no design, by accident, at random over eons of time. As if Microsoft Word was created by mixing millions of random bits until it was formed. That two entities, male and female, are perfectly matched, even though they "evolved" separately, without any knowledge of the contributions of the other entity.

Or Design

Everything was designed for a purpose by a very great designer who understands all the systems and functions. I cannot accept the belief that such elegant, purposeful design came out of random natural processes. There is a synergy between man, plants, and the cosmos. The evolutionist must take into consideration the whole package. There are dependencies that cannot be ignored. It is not just biological systems that need to be considered, but all systems. Everything is delicately balanced. We breathe in oxygen and exhale carbon dioxide (CO_2); plants take in CO_2 and give off oxygen—a perfect synergistic balance. Why? No life could exist without oxygen and CO_2. God created balance in the universe. There are many systems that are dependent on each other. This is a big problem for the global warming enthusiasts; university studies have shown that increased CO_2 has the positive effect of better plant growth and less water consumption, resulting in better and faster growth of fruits, trees, and vegetables. The Parrot fish eats algae off the coral. Every once in a while, the fish will snip off a piece of the coral. It was thought that this was damaging to the coral. The scientists roped off the coral and did not allow the Parrot fish to go near it, and the coral soon died because of the excess algae. There are far too many relationships of this type to go over. This is proof of a universal "irreducible complexity." Which came first, the coral or the Parrot fish? Both are mutually dependent.

The evolutionary story is filled with words like imagine, arose, adapted, or given rise to... (Sounds very much like a fairy tale.) But they NEVER explain what that means. I am taking a university-level course taught by a medical doctor; he said over and over again, "Look what evolution gave us." He never explained how undirected chance gave us anything or by what step-by-step process these things evolved. You are expected to accept what they say because they are the experts. You have the right, however, to question them further and ask them to explain in detail HOW it adapted or HOW it arose. What caused the adaptation and how did the organism "adapt" itself? Adaptation would have to be reflected in the DNA. How can an organism change its DNA—mutations? You will find that mutations are virtually never beneficial. Even if a mutation were beneficial, there would have to be a chain reaction of events that allows that mutation to function and be passed on to the next generation. The whole evolutionary idea is filled with speculation and guesses. Many scientists can tell you how something works, the heart, for instance. They can explain the function, why it does what it does, but they can never tell you they know HOW life began and HOW these complex functions could be developed, other than broad general statements (guesses). There are far too many interdependent functions that cannot be developed in stages. They either exist from the start, or the organism does not survive. Just having all of the chemicals together does not produce life! There must be organization, purpose, a plan, or a program that gets things moving. We will discuss these in detail as we explore the entire body.

You can believe (by faith in the absence of clear evidence) that nature rolled the dice and it came up with the right number time after time in the undirected chance process known as evolution. I choose not to believe that you can kiss a frog and a prince will appear. The atheist can never give credit to a Creator God, but they might be willing to accept an alien. How many movies and TV shows in the recent past have been made about aliens? Is this the brainwashing of people to accept the alien explanation for life on earth?

In Romans, we are told what happens to a society that rejects God. We are told that God will give them over. Romans 1:19 - 32 KJV

Because that which may be known of God is manifest in them; for God hath shewed [it] unto them. For the invisible things of him from the creation of the world are clearly seen, being understood by the things that are made, [even] his eternal power and Godhead; so that they are without excuse:

Because that, when they knew God, they glorified [him] not as God, neither were thankful; but became vain in their imaginations, and their foolish heart was darkened. ***Professing themselves to be wise, they became fools,*** *And changed the glory of the uncorruptible God into an image made like to corruptible man, and to birds, and fourfooted beasts, and creeping things.*

Wherefore God also gave them up to uncleanness through the lusts of their own hearts, to dishonour their own bodies between themselves: Who changed the truth of God into a lie, and worshipped and served the creature more than the Creator, who is blessed for ever. Amen.

For this cause God gave them up unto vile affections: for even their women did change the natural use into that which is against nature: And likewise also the men, leaving the natural use of the woman, burned in their lust one toward another; men with men working that which is unseemly, and receiving in themselves that recompence of their error which was meet.

And even as they did not like to retain God in [their] knowledge, God gave them over to a reprobate mind, to do those things which are not convenient; Being filled with all unrighteousness, fornication, wickedness, covetousness, maliciousness; full of envy, murder, debate, deceit, malignity; whisperers, Backbiters, haters of God, despiteful, proud, boasters, inventors of evil things, disobedient to parents, Without understanding, covenantbreakers, without

> *natural affection, implacable, unmerciful: Who knowing the judgment of God, that they which commit such things are worthy of death, not only do the same, but have pleasure in them that do them.*

Belief in evolution is an insult to the Creator. This belief has consequences. In 1962, using no precedence and no law to back it up, the Supreme Court of the United States officially rejected God, not by the Constitution that had governed us since our beginning and not by any law. By a stroke of the pen, one unelected body changed the destiny of this nation. Since then, we have been sliding towards the ultimate destruction of our way of life, under the blessing of God. The Supreme Court is to interpret the law; only Congress can make laws! But we find over and over that their "opinion" becomes the law of the land; this is most unfortunate!

Forty-seven years later, God has given us over to Barack Hussein Obama, who is systematically bringing further judgment on this nation by arrogantly declaring that we are no longer a Christian nation. He is systematically moving this nation towards godless socialism. They cannot be slow about it because too many people would object once they discover what he and his minions are attempting to accomplish. Evidence for this speed is having a Congress that will vote on bills they have not even read. Every day, it seems there is another attack on our God-given liberties. In recent days, states have passed laws that protect our religious liberty; the result is that organizations like the National Football League and Disney Studios have declared that they will no longer do business in those states! This government is trampling underfoot the very foundation of this once great nation: our allegiance to and trust in God. Is this the fulfillment of the judgment that Billy Graham alluded to when he said, "If God does not judge America, He will have to apologize to Sodom and Gomorrah"? We are already seeing the result of removing God from our culture. One thing is sure: if God does not bring revival to this nation in a way we have never experienced, and if we do not repent, God's judgment on our culture will be swift and complete. May God have mercy on our souls. Now in 2025 has the tide turned? We pray so...

Evolution is a stricktly faith-based model

Issue	Creation Model	Evolutionary Model
Origin of Life	**God** Genesis 1 "In the beginning God created..."	**Unsure** - maybe from a primordial soup or directed panspermia
Belief Requires	**Faith** - No observation possible	**Faith** - No observation possible
Variety of species / Kinds	**God** Genesis 1:25 "And God made the beast of the earth after his kind, and cattle after their kind, and every thing that creepeth upon the earth after his kind: and God saw that [it was] good.	Fortuitous mutations and natural selection over eons of time. Possibly punctuated equilibrium.
Belief Requires	**Current observation of kinds**	**Faith** in the face of no evidence
Evolution	**Micro**evolution = **Yes** (Variety with in kinds) **Macro**evolution = **No** (different species from others)	Micro & Macro Evolution = **Yes** One animal can become another type of animal - everything evolved from one single entity.
Belief Requires	**Current observation of kinds**	**Faith** in the face of no evidence
Life	Biogenesis = **Yes** Life only comes from other life.	Abiogenesis = **Yes** Life arises from non-life through chemical evolution How? **unsure** too many hypothesizes to list.
Belief Requires	**Current observable facts**	**Faith** in the face of no observable evidence
DNA - There are approximately 3 billion nucleotides in the DNA of a mammalian cell	**God's design** for life	**Undirected random mutations** caused the 3 million nucleotides to form in the correct order. Over eons of time.
Belief Requires	**Faith with observable facts evidence of design**	**Faith** in the face of no observable evidence

"Life is not a property of substance; it's a property of **organization**".

Dr. Gary Parker

THE BODY MAP SERIES
The BODY MAP
Guide to the Homo Sapien
Genetics
The BODY MAP
Guide to the Homo Sapien
The BODY MAP
Book III - The Five Senses
The BODY MAP
The BODY MAP
HOW I REVERSED TYPE II DIABETES
The BODY MAP
Book V - Organs Of The Body
The BODY MAP
Book VII - The Heart & Circulation
The BODY MAP

1

What is The Body Map Series?

"The body is a self-building machine; a self-stoking, self-regulating, self-repairing machine—the most marvelous and unique automatic mechanism in the universe." - Scottish biologist Sir J. Arthur Thomson.

This series is presented in multiple volumes, so you can read about the specific part of the body that you are investigating. I will attempt to be as clear as possible while divulging the necessary details without being overly technical. Some topics will be repeated in subsequent volumes with added detail.

Each book details, from what we currently know, a specific body function or system. Dr. Gary Parker makes an incredible case for design in his book. *Creation: Facts of life.*

Dr. Gary Parker explains that just as scientists understand how airplanes fly, they also recognize that flight is not a property of the materials themselves, but of how those materials are arranged and designed. A Boeing 747 consists of millions of parts that cannot fly on their own, yet through intentional design and ongoing maintenance, the plane can function and fly. In the same way, life is not simply the result of the materials that make up a cell, but of the precise and purposeful organization of those non-living molecules. A living cell is alive because of its structure and energy systems, and when those fail, life ceases and the cell breaks down. Parker emphasizes that it's not the physical components—the "dust"—that give something function or meaning, but the way they are arranged. Just as metal and glass do not inherently fly, or ink and paper do not create a novel, life itself is the result of complex organization. When that organization is lost, what remains are the basic elements lifeless and disordered.

Book 1 - Designed from the Beginning

This book is an introduction to what the Darwinists believe, plus a look at the microbiological level of the cell and its various parts and functions. The book is not intended to be a science book; however, I will provide as much technical information as needed to demonstrate the impossibility of Darwin's evolution. I hope that it raises a lot of questions in your thinking. From a computer science perspective, I will illustrate the functions with flow charts, detailing the operation and feedback loops found in every part of the body.

The evolutionists are famous for their 'matter-of-fact' reasoning. Scientists understand how the body works, but have never been able to explain how life came from non-life. They have never been able to explain, in detail, how development could occur in small steps. For the most part, the body is an all-or-nothing proposition. This series will discuss why the body could never have been developed in small increments. It is essential that all the parts are there, working in concert, or there is no life.

If you suddenly came upon a machine of unknown origin, you could certainly take it apart and examine it to explain how it works, but you could not necessarily know where it came from or how it came into being. Science is very good at discovering how our body works, making new discoveries continuously. However, they have yet to explain how it came into being or even how it "evolved." So far, their thinking has been nothing more than guesses, conjectures, and 'what if' thinking.

Book II - Reproduction

This volume will detail all of the systems and functions that must exist for human life to be created. In this volume, we will detail the female and male reproductive processes. We will show the incredible design found in the reproductive process. You will be amazed when you realize all of the parts and interactions that must take place for life to be created and what happens after birth.

The fact that there are male and female organisms is, to me, the clearest evidence for design in creation. How would two entities that depend on each other for reproduction evolve separately without a designer? They did not evolve and could not have evolved, as the Darwinian evolutionists suggest. There is no time for testing different approaches through natural selection because the process has to work the first time, or life stops with that generation.

> How is it possible that embryonic development does not entail a disorderly growth of cells, but is systematic and purposeful according to a set timetable? A precise plan, in which all stages are programmed in

> the finest detail, underlies all these processes. In this case also, information is the overall guiding factor.
> Dr. Werner Gitt - *In the Beginning was Information.*

Book III - The Five Senses

Sight, touch, hearing, smell, taste. How they function and what has to happen for them to be useful to the body.

We are told that evolution accounts for all living systems, but we are not told exactly how. Tell me, in intricate detail, how the eye could evolve step by step. This is the trick of the evolutionist's mindset; they will make claims about what may have occurred. With great confidence, they declare evolution as fact. Where is the evidence? Ask an evolutionist how the first life evolved and continued to evolve into ever more complexity and get ready for a "why" or "what could have happened," but never "how." Richard Dawkins has the answer in his book, Climbing Mount Improbable (W. H. Norton & Company, Inc. 1996).

> "...acts by breaking the improbability up into small manageable parts, smearing out the luck needed, going round the back of Mount Improbable and crawling up the gentle slopes...."

To me, this sounds like a lot of psychobabble and wishful thinking. He still does not explain how an extremely complex structure like the human eye can evolve on its own, undirected. Where does the new information come from that directs these small steps? New information must be coded in the DNA to be passed on. How does that happen? Mutations? Current observations and testing prove that mutations do not increase information. In fact, we have observed that all mutations are virtually always harmful. Natural selection would not "freeze" these new features unless they were in some way beneficial according to its hypothesis. It is still speculation, conjecture, and there is no evidence at all for evolution. It is wishful thinking on Dawkins' part to hope that these small parts would end up doing anything but causing trouble.

The eye is not just a photoreceptor; it is a complex organ that must have connections to the brain to work. How did that section of the brain evolve simultaneously? It is a complete system or nothing. David Berlinski nails the issue to the wall in rebuttal to Dawkins' ideas of Mount Probable.

> The very problem that Darwin's theory was designed to evade now reappears. Like vibrations passing through a spider's web, changes to any part of the eye, if they are to improve vision, must bring about changes throughout the optical system. Without a correlative increase in the size and complexity of the optic nerve, an increase in the number of photoreceptive membranes can have no effect. A change in the optic nerve must in turn induce corresponding neurological changes in the brain. If these changes come about simultaneously, it makes no sense to talk of a gradual ascent of Mount Improbable. If they do not come about simultaneously, it is not clear why they should come about at all. *Keeping an Eye on Evolution:* David Berlinski R*ichard Dawkins, a relentless Darwinian spear carrier, trips over Mount Improbabl*e. The Globe & Mail, November 2, 1996, p. D10

Book IV - The Brain and The Nervous System

The human brain and the nervous system exemplify an extraordinary level of complexity, which guarantees a deep exploration of their design and functionalities. The brain, a highly organized network of neurons, greatly exceeds any computational system made by man, demonstrating an architecture that operates and processes information in a way that is not yet completely understood. Together with the nervous system, the brain is responsible not only for the basic physiological functions but also for the complex tapestry of human behavior, cognition, and emotions. From our perspective, this remarkable design suggests an intelligent origin, which contrasts strongly with the notion of random evolution.

In this book, we will explore the wonders of the nervous system through the lens of creation science. You will encounter the brain not as an evolutionary accident but as a crafted interface between body and soul, a biological marvel that supports thought, emotion, memory, creativity, and even spiritual communion.

> The nervous system is a biological system that regulates the body's responses to internal and external stimuli, by carrying information to and from all parts of the body through a vast network of nerve cells. Nervous tissue is found in the brain, spinal cord, and nerves. It is responsible for coordinating and controlling many body activities. It stimulates muscle contraction, creates an awareness of the environment, and plays a major role in emotions, memory, and reasoning. To do all these things, cells in nervous tissue need to be able to communicate with each other by way of electrical nerve impulses - *Nervous Tissue SEER Training Modules, National Cancer Institute.*

Book V - Organs

This volume covers the major organs of the body, with the skin being the largest. The problem for the evolutionist is that the body cannot function without these major organs all being in place at the same time. Small incremental steps will not work.

In this book, we will attempt to show the complexity and interoperability of the major organs of the body and how they function.

Book VI - The Endocrine Systems

Without the endocrine system, you would not exist! Evolution can never explain how humans could exist without a fully functional endocrine system. There are so many processes that are interdependent that missing one function would cause sickness and death before a step-by-step evolution could take place. Without a fully functional endocrine system, you would not be able to procreate, regulate sugar levels, sleep, among many other critical functions. The evolutionist should reread Darwin:

"If it could be demonstrated that any complex organ existed, which could not possibly have been formed by numerous, successive, slight modifications, my theory would absolutely break down."
The Origin of Species - Charles Darwin

Darwin had no idea in his day... When you see how complex the endocrine system is and how it affects various parts of the body, you will see how insane the idea of evolution is when explaining the origin of life.

Geoffrey Simmons, M.D., adds a new term, "whole package phenomenon," in explaining the interrelationship of hormones in human physiology. The endocrine system is made up of major glands such as the pituitary, thyroid, adrenal glands, as well as the testes and ovaries, along with smaller hormone-producing tissues found in organs like the pancreas, heart, lungs, kidneys, stomach, liver, and placenta. Each of these glands releases specific hormones that act as messengers, directing target cells on how to function and when to respond. Scientists have identified over 40 hormones so far, each regulated through feedback mechanisms. Some hormones function together, others counteract each other, and some are controlled by multiple feedback systems. This high level of interdependence and the necessity for all parts to work together from the beginning—the "whole package phenomenon"—poses a significant challenge to evolutionary explanations.
—Geoffrey Simmons, M.D., *What Darwin Didn't Know* (p. 167)

Just like the eye and many other critical functions in the body, the immune system depends on several systems to operate: the thymus, spleen, lymphatic system, bone marrow, white blood cells, antibodies, and hormones, just to name a few. Our description of the immune system in this book only introduces you to the many things that go on in the immune function of our bodies. The evolutionist has a difficult task ahead: explaining how this marvel of intricate and synergistic design came about through undirected chance mutations. For the immune system to work properly, all of the pieces had to be there from the beginning; otherwise, no human would survive the relentless attacks of foreign substances entering our body every minute of the day. Without a complete immune system, you would be sending an army to stop invaders, but they don't know who they are, what they look like, or how to defeat them. They would have no direction and no weapons to work with and would soon be overtaken by the enemy. No! Our bodies would not have survived without the marvel of the immune system.

Book VII Circulatory System and the Heart

Another critical system is how to get nutrients, oxygen, and other vital elements to each cell and carry away waste and deoxygenated blood. What is flowing through your veins may surprise you. When you discover what it takes to clot blood when you get a cut, it will amaze you. God, who created life, knew from the beginning the importance of blood. The idea that the Bible has no scientific significance is foolishness.

For the life of the flesh is in the blood: Leviticus 17:11

Book VIII Lymphatic System

The lymphatic system is a large part of the immune system and serves as a defense against disease. Fluids are passed through a network of lymph capillaries and are pushed through one-way valves by the body's muscles.

> The defense department of the body's immune system headquarters in bone marrow, which continuously produces cells that are processed into three specialized divisions. One division becomes promonocytes, then monocytes the potential attack force. Altered by the thymus gland, a second division becomes T cells, the signal corps. The third division becomes B cells, the supply corps. These three divisions make a highly mobile, coordinated brigade capable of detecting and destroying an invader. "Bacteria invade, perhaps through a tiny scratch, and begin to multiply. Some become stuck to monocytes, activating [changing] them into more aggressive macrophages. Circulating randomly, T cells adhere to macrophages and to

the bacteria. This linkage stimulates T cells to send three messages. One: Activate and summon more macrophages. Two: Produce more T cells. Three: Signal B cells to proliferate and become plasma cells that produce and release antibodies Y-shaped molecules. Antibodies latch onto intruder molecules, or antigens, on bacteria and link the yoke of the Y to them. The shaft of the Y attaches to macrophages. They, in turn, envelop the bacteria, which are consumed by enzymes and become harmless debris. Rick Gore, *Awesome Worlds Within a Cell, National Geographic Society,* September 1976, pp. 378-379.

Book IX Respiratory System and the Lungs

To believe that a system could evolve in small steps is absurd. Which came first, the heart or the lungs? You cannot live without both present at the same time. The heart delivers oxygen to the cells and brings back CO2. I cannot accept that blind, undirected chance could evolve that function.

It is amazing that we take in oxygen and breathe out CO2, and that plants take in CO2 and give off oxygen. This synergy cannot be explained by undirected chance. How and why would undirected chance create one thing for prokaryotes (plants) and another for eukaryotic systems (animals)?

Book X Digestive System

The human digestive system is a marvel of intricate design and purposeful function. From the moment food enters the mouth to the final stage of nutrient absorption and waste elimination, each organ works in perfect harmony to sustain life. This system is not a product of random mutations or evolutionary trial and error—it bears the unmistakable signature of a wise and intentional Creator.

Every component, from the salivary glands to the stomach, intestines, liver, and pancreas, is fine-tuned to break down food, extract vital nutrients, and protect the body from harmful substances. The coordination required between muscular contractions, enzyme release, and chemical balances reveals an astounding level of complexity that defies chance.

In this book, we will explore the digestive system not only as a biological mechanism but as a testimony to the craftsmanship of our Creator—one who designed our bodies with care, foresight, and purpose.

Book XI Skeletal System & Muscular System

The skeletal and muscular systems form the structural and functional foundation of the human body—an integrated design that reflects order, strength, and purposeful motion. These systems are not the result of blind evolutionary processes but rather the intelligent handiwork of a Creator who fashioned the human body with precision and care.

The skeletal system provides the framework that supports and protects vital organs, anchors muscles, and enables upright posture. Bones are not inert scaffolding; they are living tissues capable of repair, regeneration, and even producing blood cells—evidence of intentional complexity.

Working in perfect unity with the skeleton, the muscular system allows for movement, stability, and the regulation of vital functions such as breathing and circulation. Each muscle is uniquely shaped and placed for a specific task, whether it's lifting an arm, running, or maintaining posture during stillness. This synergy between bones and muscles, guided by the nervous system, is a masterpiece of biomechanical engineering.

In this book, we will examine the skeletal and muscular systems as a testimony to God's wisdom—systems that not only allow us to move and live but also reveal the glory of the One who designed us "fearfully and wonderfully" (Psalm 139:14).

Book XII Final word on Creation

The human body is a breathtaking masterpiece of design, coordination, and purpose. Every cell, tissue, and organ works in harmony to sustain life, adapt to change, and perform functions far too complex to be the result of chance. From the beating of the heart to the firing of neurons in the brain, from the healing of wounds to the digestion of food—our bodies display systems of interdependence and precision that declare the handiwork of an intelligent Creator.

Rather than being the product of random mutations and undirected processes over millions of years, the human body reveals the mark of intentionality and foresight. The skeletal and muscular systems give us strength and mobility. The circulatory and respiratory systems ensure that oxygen and nutrients are delivered exactly where they are needed. The nervous system coordinates it all with lightning speed, while the immune system stands guard, protecting us from harm. Even our reproductive system is designed with incredible intricacy to pass on life and genetic information in an orderly fashion.

No machine, no invention of man, comes close to matching the complexity, adaptability, and beauty of the human body. It is not merely a biological organism; it is a living testimony to the Creator who formed us in His image. As we journey through each system in this book, we will not only study the anatomy and physiology of the body—we will also uncover the undeniable evidence of intelligent design, purpose, and divine craftsmanship.

This book will conclude the series and illustrate a step-by-step look at how the body parts operate.

DISCOVERING
THE ORIGIN
OF LIFE
BODY
MAP

THE ORIGIN OF LIFE

2

The Origin of Life

As thou knowest not what is the way of the spirit, nor how the bones do grow in the womb of her that is with child: ***even so thou knowest not the works of God who maketh all.*** Ecclesiastes 11:5 (KJV)

The followers of Darwin's ideas believe that life sprang from non-life. This is known as abiogenesis (Greek a "without," bios "life," and genesis "beginning, origin"), the idea that life comes from non-life. Even Crick, co-discoverer of the DNA helix, realized that it would take a "miracle" for life to come from non-life. Dr. Crick calls it "almost a miracle." Miracles do not exist in random, non-directed processes, do they?

> *What is so frustrating for our present purpose is that it seems almost impossible to give any numerical value to the probability of what seems a rather unlikely sequence of events... An honest man, armed with all the knowledge available to us now, could only state that in some sense, the origin of life appears, at the moment to be almost a miracle.*
> *Dr. Francis H. Crick, in his 1981 book, Life Itself*

Life is the result of design and organization. Many authors have used the airplane as an example. The airplane is made up of thousands of non-flying parts. Any one part cannot fly on its own. But combine the parts, in an organized manner, with a design that is meant to fly, and you have a machine that can fly. Similarly, your body is made up of billions of non-living parts. Alone, any one of the parts is not alive, but when put together, in an organized manner, after a specific design, you have life.

Dr. Jerry Bergman underscores a fundamental challenge to naturalistic explanations for the origin of life: the necessity of simultaneous functionality. He points out that biological components cannot evolve gradually and independently because isolated parts of a living system quickly degrade in the absence of a complete and functioning whole. Without mechanisms like membranes to protect them, enzymes to sustain reactions, and DNA to provide instructions, essential biomolecules would succumb to environmental damage—such as dehydration, oxidation, and microbial attack. This reality strongly supports the idea that life must have originated as a fully integrated system, not through slow, stepwise evolution. Bergman reinforces this by referencing evidence such as the rapid degradation of nucleotides under early Earth conditions, which poses a serious problem for theories of abiogenesis. § From a creation science perspective, his conclusion aligns with the biblical view that life began through the instantaneous and intelligent act of creation, not by chance-based processes over vast ages.

Just having the parts will not spark life. But when molded by design and organized in a specific way, life is possible. Even if the evolutionist had all of the building blocks of life in their "primordial soup," it would not create life. If you believe having all of the necessary chemicals will create life, then why can't we reanimate the dead, like Frankenstein's monster? There is more to life than just chemistry.

If you were to build a plane, where would you start? First, you would need the design—a blueprint explaining what is needed and how the parts should come together. Assuming that you can get all of the parts, someone would need to assemble them. Assuming you could assemble it, you would need fuel to make it go. Once you get the plane off the ground, there are numerous feedback loops that will inform the pilot about the state of the flight: airspeed, altitude, and fuel remaining, among many other things. If you could get life from non-life by accident, where are the feedback loops to maintain life and fuel processing to sustain it? What about defense against harmful intruders? Where is the intelligence that tells the cell when it needs to divide and perpetuate itself? Any one of these issues would eliminate an accidental cell from surviving very long. That does not take into consideration the harmful environment in which it may find itself fighting for life.

> If you can't explain where information comes from, you haven't explained life, because it's the information that makes the molecules into something that actually functions. - Stephen C. Meyer - *Case for the Creator p225 by Lee Strobel*

The issue for abiogenesis:

To make even the simplest living organism requires design, organization, fuel, and reproductive capabilities. The design comes in the form of DNA. However, to get the simplest cell, one requires proteins and much, much more.

> So far, no geochemical evidence for the existence of a prebiotic soup has been published. Indeed, a number of scientists have challenged the prebiotic soup concept, noting that even if it existed, the concentration of organic building blocks in it would have been too small to be meaningful for prebiotic evolution. - Dr. Noam Lavav *Biogenesis: Theories of Life's Origins,*

> The major problem is the origin of the genetic code and of its translation mechanism. Indeed, instead of a problem, it ought rather be called a riddle. The code is meaningless unless translated. The modern cell's translating machinery consists of at least fifty macromolecular components which are themselves coded in DNA: the code cannot be translated otherwise than by products of translation. It is the modern expression of omne vivum ex ovo [everything that lives, (comes) from an egg]. When and how did this circle become closed? It is exceedingly difficult to imagine. - Jacques Monod, *"Chance and Necessity: An Essay on the Natural Philosophy of Modern Biology"*

Getting all the right ingredients for life

Amino Acids

It is the chains of amino acids that make up proteins, which, in turn, make up living things. There are natural chemical reactions that take place. Water is one of the best and most widely used solvents. For instance, water dissolves salt.

It is water's chemical composition and physical attributes that make it such an excellent solvent. Water molecules have a polar arrangement of oxygen and hydrogen atoms—one side (hydrogen) has a positive electrical charge, and the other side (oxygen) has a negative charge. This allows the water molecule to become attracted to many different types of molecules. Water can become so heavily attracted to a different compound, like salt (NaCl), that it can disrupt the attractive forces that hold the sodium and chloride ions in the salt compound together and, thus, dissolve it. - Http://water.usgs.gov/edu/solvent.html

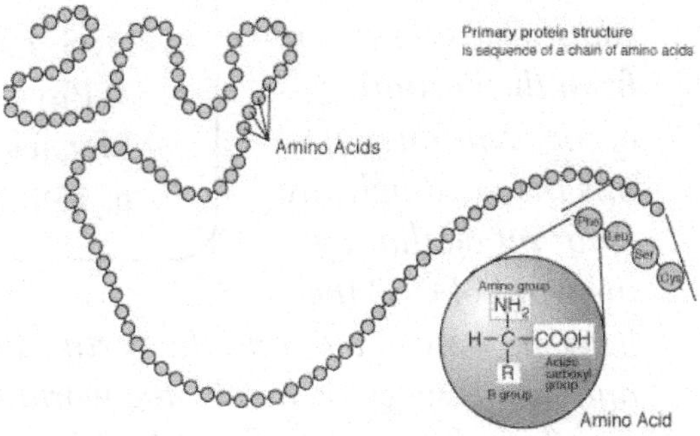

Public Domain from the
National Human Genome Research Institute.

However, the chains of amino acids that make up proteins are linked together in a very specific way, using only left-handed amino acids. There is virtually no natural attraction of amino acids that would form a specific chain of protein. It is your DNA that has the instructions for making the proteins necessary for life. There is a trend among evolutionists to abandon the idea of abiogenesis (in the general theory of evolution) and instead focus on the diversification of life (the special theory of evolution). If they cannot explain how life started through undirected natural processes, they cannot explain how things began without a designer. This, then, makes the diversification of life a matter of the outcome of design as well. The process had its origin with a designer. All efforts to describe how life began through natural processes have failed. If life did not start with natural, undirected processes without a Creator, why would life need evolution to continue the process? God created everything in six days and said it was very good.

The god of the theistic evolutionists must not be very powerful if they say God started it but then evolution took over. They take the verse in Peter out of context, applying it to mean that God could have taken millions of years to get everything we have today."

Knowing this first, that there shall come in the last days

scoffers, walking after their own lusts, And saying, Where is the promise of his coming? for since the fathers fell asleep, all things continue as they were from the beginning of the creation. For this they willingly are ignorant of, that by the word of God the heavens were of old, and the earth standing out of the water and in the water: Whereby the world that then was, being overflowed with water, perished: But the heavens and the earth, which are now, by the same word are kept in store, reserved unto fire against the day of judgment and perdition of ungodly men. ***But, beloved, be not ignorant of this one thing, that one day is with the Lord as a thousand years, and a thousand years as one day.*** *The Lord is not slack concerning his promise, as some men count slackness; but is longsuffering to us- ward, not willing that any should perish, but that all should come to repentance.* 2 Peter 3:3-9

> **"...evolution became in a sense a scientific religion; almost all scientists have accepted it and many are prepared to 'bend' their observations to fit with it."**
>
> **H. S. Lipson,**
> **A Physicist Looks at Evolution,**
> **Physics Bulletin,**
> **p. 138, May 1980**

This verse has nothing to do with the length of time of creation. But, as is always the case, you can make the scripture say anything you want if you twist it and take it out of context. Or, as Chuck Missler often says, "If you torture the data long enough, it will confess to anything."

I do not buy that idea. My God is great and powerful and can, by the word of His mouth, create everything in a literal six 24-hour periods. For further proof, I recommend the book *In Six Days: Why Fifty Scientists Choose to Believe in Creation by John F. Ashton (Editor).*

Cell Membranes

Cells cannot exist without an outer membrane to protect the inner contents. There must be a way to let nutrients in but keep invaders and inappropriate chemicals out. It is also critical to release waste materials. The membrane is not just some wrapper like plastic wrap. Even the simplest of cells are far from being simple. Read what Arthur V. Chadwick, Ph.D. has to say about the membrane.

Dr. Arthur V. Chadwick, (*Abiogenic Origin of Life: A Theory in Crisis)* in his critique of the abiogenic origin of life, highlights the immense complexity and challenges involved in forming even the most basic element of life—a functional cell membrane. While it is true that lipid bilayers can spontaneously form spherical, cell-like structures under certain conditions, the actual requirements for a living cell are far more demanding. All living cells need a membrane composed of phospholipids, stabilized and structured by specialized proteins. However, the formation of fatty acids—essential components of these membranes—has proven exceedingly difficult under supposed prebiotic conditions, even when assuming a reducing atmosphere. Moreover, naturally occurring elements like magnesium and calcium would quickly bind to these fatty acids, causing them to precipitate and settle into ancient sediments, rendering them unavailable for membrane formation. Even if a primitive membrane could somehow form, it would be impermeable to many of the molecules a cell needs to survive. Modern cells solve this with highly specialized membrane proteins that regulate molecular entry, but such proteins would not exist in a hypothesized "first protocell." This creates a paradox: a membrane is essential for life, yet its very presence without the supporting structures would prevent life from emerging. Thus, the spontaneous development of a cell membrane—a prerequisite for life—is far from assured and stands as a serious challenge to abiogenesis theories.

Proteins

When you eat protein, it enters the stomach, where it is broken down into amino acids. (Appendix C has a list of these amino acids.) These are categorized as essential and non-essential. The essential amino acids are those that are not produced in the body itself, while the non-essential ones are created in the body. The small intestine then absorbs the essential amino acids into the bloodstream via the lacteal villi and passes them into the bloodstream. This is then processed in the liver and moved to the cells for further processing into a final product of various proteins.

We will discuss how proteins are made later. What we know for certain is that even the simplest single-celled organisms have a multitude of very specific proteins and functions more complex than one can imagine. Evolutionists cannot explain how these proteins came together without instructions. However, even if one had all of the right chemicals and even the instructions to put them together, one still needs someone or something to build it. Random processes would never be able to do it on their own. It is pure wishful thinking to assume it would. Even the famous Miller-Urey experiment (described earlier) only proved that it could not be done.

The problem for evolutionists increases when one factors in the issue of reproduction. Even if one has all the right chemicals and can build it, how does the entity reproduce itself? How would it know to reproduce? Where would the information and the mechanism come from to reproduce? Blind chance could NEVER evolve two entities that work perfectly together, male and female, each bringing something to the reproductive process that the other does not have!

How do you ensure that the fuel needed for survival exists and is in a form that can be used? Darwin had no clue about the operation of a cell. It is an all-or-nothing scenario. Far too many parts have to be there at the beginning, or you have no beginning. All attempts to prove that random undirected chance could build the first living cell have failed. If Darwinists cannot explain how life got started, how can we believe the rest of the fairy tale? Scientists can tell us how the body operates, but they have never explained how it all came together without a Creator/designer. Too many systems are interdependent and therefore must all be in place at once. Darwin's idea that things evolved over eons of time, in small successive steps, is false. The various university courses I have listened to make the same false premise: "this function evolved to give us X." However, they never explain in detail how it evolved. This is something that students must ask: "How did it evolve? What processes caused it to evolve? What is the step-by-step process? It is easy to explain how something works, but it does not explain how it got that way.

Enzymes

Virtually all enzymes are proteins. Without enzymes, life would not be possible. Enzymes speed up the processing of reactions in the cell, and they are very specific to the chemicals they react with, much like a lock and a key. Wrong key, no reaction, no life. Additionally, enzymes are very sensitive to temperature, pH, and other factors. This is another reason the primordial soup could not produce life: enzymes would not survive the harsh environment.

Enzymes speed up this process more than a million times. Without enzymes to speed things up, life would not exist because the reaction time would be too long, and life would stop. Here is yet another example of why the body is an all-or-nothing proposition. Without enzymes in the cell from the beginning, there would be no start and certainly no sustaining life. These enzymes are very specific and carry out very specific tasks. Where did the information come from to build these wonders of the cell? Random chance and eons of time cannot explain it. Life requires design and organization to exist.

What are the minimum requirements for a living cell?

1. A cell membrane to protect the contents from the outside invaders.
2. Obviously, you must have the instructions for making proteins, DNA.
3. The proteins and functions necessary to read the instructions to build the proteins. The DNA must be unwound and read and mRNA created. (More detail on this, later in the DNA chapter)
4. It takes information.
5. A way to take in nutrients and process them

6. A way to generate energy (ATP)
7. A way to reproduce itself, just to name a few things required.

Information

> The way life manages information involves a logical structure that differs fundamentally from mere complex chemistry. Therefore, chemistry alone, will not explain life's origin, any more than a study of silicon, copper and plastic will explain how a computer can execute a program. Our work suggests that the answer will come from taking information seriously as a physical agency, with its own dynamics and causal relationships existing alongside those of the matter that embodies it – and that life's origin can ultimately be explained by importing the language and concepts of biology into physics and chemistry, rather than the other way round. *The secret of life won't be cooked up in a chemistry lab* - Paul Davies Http://www.theguardian.com/commentisfree/2013/jan/13/secret-life-unveiled-chemistry-lab

Life cannot exist without information and organization (programmed purpose). Blind chance can never account for the absolute precise chemicals coming together at the right time in the right way. Darwin was right when he said that his theory would fall apart if it could be shown that life could not have evolved in minute steps.

In *The Meaning of Evolution*, Dr. George Gaylord Simpson emphasizes the extraordinary complexity found even in the simplest forms of life. He challenges the often-repeated idea that evolution progressed smoothly "from amoeba to man," as if the amoeba represented a primitive, uncomplicated starting point. In reality, Simpson points out that the leap from non-living molecular structures to a fully functional single-celled organism like an amoeba is at least as vast—and perhaps more significant—than the evolutionary distance from an amoeba to a human being. The single-celled protozoan already possesses solutions to all the fundamental challenges of life: metabolism, reproduction, adaptation, and environmental interaction. These solutions are not invented anew in higher organisms, but merely elaborated upon. Thus, the origin of even the "simplest" life form represents a staggering hurdle, one that underscores the depth of complexity present from life's very beginning.

As we investigate the entire body, you will find that information is a critical component of life as we know it. There are processes that cannot work without the information needed to direct their operation. The immune system has to know when it has been invaded by a foreign entity, such as germs and viruses. The function of keeping the body in balance through homeostasis requires a massive amount of information, such as body temperature, blood sugar levels, and others. The brain has to know how to interpret sight and sound, to name just two. None of this information could have come from evolution.

William Dembski tells us that DNA is both complex and specified simultaneously and, therefore, it must have been produced by an intelligent cause, rather than being the result of natural processes."*Intelligent Design as a Theory of Information Access Research Network, November 15, 1998*

In his explanation of specified complexity, Dr. William Dembski illustrates a critical distinction in understanding information. He notes that a single letter of the alphabet is specified—that is, it has meaning or identity—but it is not complex. Conversely, a long sentence composed of random letters may be complex, showing no discernible pattern or order, but it lacks specification—it carries no intended meaning. However, a Shakespearean sonnet is both complex and specified; it displays intricate structure and conveys purposeful content. This combination, according to Dembski, is a hallmark of intelligent design, as such dual qualities are not known to arise through unguided, random processes.

It is very interesting that the entire blueprint for life (DNA) only contains four letters in its definition (five if you count RNA, which replaces T with U). The syntax that defines an amino acid used in building proteins and enzymes is represented by a three-letter codon. Within the DNA are other letter combinations that define many other functions, such as start and stop points, as well as many things that are still not understood.

> Whatever else a living creature may be...[it] is also a combinatorial system, its organization controlled by a strange, a hidden and obscure text, one written in a biochemical code. It is an algorithm that lies at the humming heart of life, ferrying information from one set of symbols (the nucleic acids) to another (the proteins) David Berlinski, *The Deniable Darwin and Other Essays. Seattle: Discovery Institute Press.*

In *The Fifth Miracle,* physicist and cosmologist Dr. Paul Davies candidly reflects on his journey researching the origin of life. Initially, he believed that science was on the verge of solving the mystery of how life began. However, after spending a year or two immersed in the field, he came to a very different conclusion: there remains a vast and unresolved gap in our understanding. This gap is not just a matter of missing technical details, but a profound conceptual void. Davies reveals that while many scientists privately acknowledge their bafflement over the origin of life, they are often reluctant to express this publicly. He attributes this hesitation to two main concerns: first, the fear that admitting such uncertainty may give legitimacy to religious or supernatural explanations; and second, the concern that transparency about these unresolved issues could jeopardize research funding. His honesty exposes the tension between scientific curiosity and institutional pressures, highlighting the depth of the mystery still surrounding life's beginnings.

In *Evolution: A Theory in Crisis*, Dr. Michael Denton uses a vivid analogy to help readers comprehend the astonishing complexity of the cell as revealed by molecular biology. He suggests magnifying a single cell a billion times until it spans twenty kilometers in diameter—large enough to overshadow a city like London or New York. At that scale, the cell would resemble a colossal, intricately designed airship, with millions of openings on its surface functioning like port holes, regulating the constant flow of materials in and out. Upon entering, one would encounter a world of unimaginable technology and complexity—far beyond anything produced by human ingenuity. Denton challenges the plausibility that such a reality could arise from random processes alone. He argues that even the smallest functional units of life, such as a protein or gene, are complex beyond our own creative capacities. Such intricacy, precision, and order, he concludes, are not the products of chance but reflect design that vastly surpasses anything the human mind has conceived.

Michael Denton, Evolution: A Theory in Crisis,
Burnett Books, London, 1985, pp. 328, 342.

See appendix D - The Marvel of Microcosm: Unveiling the Intelligent Design and Complexity of Human Cell Structure and Function

EVOLUTION: A THEORY IN CRISIS

THE BLUE PRINT FOR LIFE

3

DNA
The Blueprint For Life

In this chapter:

What is DNA?
Is there a lot of left over junk from evolution in DNA?
How are proteins made?
What is Mitochondrial Eve?
What are the DNA doctors?
What is epigenetics?
What is a telomere?

The following are quotes from Origin of life an explanation of what is needed for abiogenesis (or biopoiesis) by Don Batten Http://creation.com/origin-of-life

"Anyone who tells you that he or she knows how life started on earth some 3.4 billion years ago is a fool or a knave. Nobody knows."— Professor Stuart Kauffman,
At Home in the Universe: The Search for the Laws of Self-Organization and Complexity, Oxford University Press, p. 31, 1995

"…we must concede that there are presently no detailed Darwinian accounts of the evolution of any biochemical or cellular system, only a variety of wishful speculations." —Franklin M. Harold *The way of the cell: molecules, organisms and the order of life,* Oxford University. Press, New York, p. 205

"The biggest gap in evolutionary theory remains the origin of life itself… the gap between such a collection of molecules [amino acids and RNA] and even the most primitive cell remains enormous." —Chris Wills, professor of biology at the University of California

"Conclusion: The scientific problem of the origin of life can be characterized as the problem of finding the chemical mechanism that led all the way from the inception of the first autocatalytic reproduction cycle to the last common ancestor. All present theories fall far short of this task. While we still do not understand this mechanism, we now have a grasp of the magnitude of the problem." G. Watchershauser, *Origin of life: RNA world versus autocatalytic anabolism, The Prokaryotes*, Vol. 1, 3rd Edition, chapter 1.11, pp. 275–283, p. 282

Intelligence and DNA

My question to the evolutionist is: if it were possible for DNA to come about by pure chance (which in itself is statistically impossible), all you have is an extremely complex book of instructions. Just having instructions (DNA) will do nothing to build proteins unless you have something to act on those instructions. Each cell can be compared to a large manufacturing plant. You have materials coming in and going out, and a set of instructions to carry out in building the product. But just having the instructions on how to build something (DNA) does not cause it to be built. What is the "intelligence or labor force" that drives these instructions to do their work? Where did this force originate? A new discovery is making the DNA even more miraculous; this process is known as epigenetics. Epigenetics makes it possible for heart cells to always make heart cells and not nerve cells or any other type of cell. Parts of the DNA that do not code for heart cells are permanently turned off in the heart cell's DNA. We will discuss in more detail how epigenetics works later in this chapter. I would guess that no person understands it fully, and yet we have a lot of information about what goes on in a cell. Darwin knew nothing about this when he formed his ideas. DNA is the full set of instructions in the genes to build a complete organism. For the organism to be built, it requires many other parts and mechanisms to keep it going. DNA is required to build proteins, and there are proteins that carry out various functions in the cell. If the process that creates new proteins from "instructions" in the DNA is in question, where did the first proteins and enzymes come from that do the job of translation? You have a chicken-and-egg problem here. To build proteins, you have to have DNA "instructions" to direct the building of a specific protein, but for proteins to be built, you have to have specialized proteins to carry out the work of building new proteins. Several factors known as transcription factors carry out the transcription process in conjunction with RNA polymerase and a whole lot more, which we will discuss later.

It is one thing to know what happens in a cell; science has a fair idea of what is happening there, and researchers are making new discoveries all the time. However, it is another thing to understand why it does what it does and why it came into being in the first place. It is similar to computer hardware: you can know everything about the hardware, but you won't know how things get started and how they will react because software drives it. There is an operating system, which we know very little about, that drives the hardware. We know, for instance, how blood clots, but the extreme complexity of the reactions that take place to make it happen is spectacular! The fact that blood does not clot until it is needed on the surface of the wound and does not cascade and clot everywhere is a miracle. I believe there is more to the cell than just chemical reactions. Consider this for a moment: when producing a new protein molecule from DNA, the following questions need to be asked:

1. Where does the intelligence come from that determines the need to make a new protein? It's not DNA. What triggers it? Who ordered it? Is it always hormones? It sounds like an operating system is at work directing things, or at the very least, a very complex algorithm programmed by a "designer"!

2. What is the "Intelligence" or program that drives transcription (the process of opening a section of DNA for copying)?

3. How does it (the "Intelligence") know which section to open and copy? Who told it what protein to build?

4. What is the driving force behind the creation of Messenger RNA (the copy of the DNA section)? Why does it delete the Introns and splice the Exons when making mRNA? How does epigenetics determine what to turn on or off?

5. What is the "Intelligence" that tells the molecular machine to move from the cell nucleus through the Nucleus Pore Complex to the Ribosome? How does it know to open the pore complex to let the mRNA through?

6. What "Intelligence" causes the Ribosome to translate the Messenger RNA into a new sequence of amino acids?

7. What "work FORCE" causes tRNA to attract the appropriate amino acids and then deliver them to the ribosome in the right sequence?

There is something that is guiding and directing the entire process and it is not blind chance! There is an operating system (intelligence) and very specific algorithms that guide and direct the entire process. Chance can never have put together such a highly programmed system that virtually works without fail 24 hours a day, through your entire life.

When I was young, I used to love putting together model cars. How long would it take for the model to be assembled if I waited for the instructions to put it together? You say, that is silly, the instructions will never assemble the model, you must follow the instructions yourself to build it. My point exactly, there must be a living "Intelligence" or program that is controlling everything. Without this "Intelligence" the cell would soon die. If you could put all the elements of a cell in a flask at the right temperature in the right solution with all the amino acids, proteins and a copy of the DNA it would still die, why? Because it is not about chemistry, it is about design and the operating system that guide the processes.

The cell is full of complex factory operations, taking place 24 hours a day, seven days a week. There are "molecular machines" that carry out the work in the cell. How do these machines know what to do? Is it pure chemical reactions? My guess is that there is something we still do not understand how and why a cell does what it does. It is as if someone was directing operations not unlike a computer operating system. It is the software that makes it do what it does. This operation could NEVER have come from undirected chance no matter how much time you have! God, in His infinite wisdom, designed a system which would self-perpetuate.

If you wanted to create a new function or even a new system, you would have to have the "information" necessary to create that new function or system, and that would need to be coded in the DNA. This "intelligence" would have to be aware that a new protein is needed. Question: where would this new information come from, and how would it get coded in the DNA? Following a careful scientific analysis, based on what we can and have observed, mutations do not create new systems or functions; they only disrupt or cripple existing ones, like a programmer who makes changes to a program without understanding what effect it will have on the entire process. And so, the question stands: where would this new information originate? All proteins are made from 20 amino acids. These must be sequenced and folded in a very precise manner, or we have a problem, sickness, or even death. Where did the knowledge originate? Blind chance? Not a chance!

However, if you involve genetic engineering to create change, then it can no longer be called evolution since it was designed by a Creator (a geneticist).

I can tell you from experience that adding a function to a program is, many times, not an easy thing to do. What typically happens is that by adding or fixing one issue, you usually end up causing other processes to fail or act in a way not intended. This is especially true in complex programs with a lot of inter-dependability. Even our most complex programs are nowhere near as complex and interrelated as the functions of the body. So, random chance or mutations would destroy other functions because they are blind to the operation of other parts of the system. To think that is possible is ludicrous!

If you watch any TV, you will see a lot of commercials for a variety of drugs. It strikes me as odd that the side effects sound worse than the problem they are trying to solve. You can tell from the commercials that the drugs, similar to modifying a program's code, can do more harm and disrupt the process more than before taking the drug. With all of the knowledge, understanding, and research that the drug companies have, they still don't always get things right and, in some cases, make things worse. Look at the numerous lawsuits against drug companies. If man creates faulty drugs through their research and knowledge, how could blind, unintelligent chance make improvements in any system? It is an absurd belief.

Can you take a program designed for Windows 10 and run it on Microsoft DOS 2.0? Certainly not! Why? Because DOS was not aware of the many things taken for granted in a Windows operating system, which is many times more complicated. Can you run a DOS program under Windows without special software? No. This may be a bad example, but it makes my point. Undirected chance would never have the intelligence to know the impact it may have on any change made by random, undirected chance! This is the end of evolution. No small changes can account for the design and complexity we see all around us. If you could get a positive mutation, it would have to change the DNA in the reproductive cell so it would get passed along. How would that happen?

During Darwin's time, very little was known about the cell. Ernst Haeckel, who promoted and popularized Darwinism in Germany, said this about the "simple" nature of the cell:

> "Homogeneous globules of plasm that were composed chiefly of carbon with an admixture of hydrogen, nitrogen, and sulfur. These parts properly united produce the soul and body of the animated world, and suitably nursed became man. With this single argument the mystery of the universe is explained, the Deity annulled, and a new era of infinite knowledge" ushered in" Haeckel (1905), *The Wonders of Life*, (trans. J. McCabe) p. 111. [my emphasis]

(He is also infamous for using fraudulent drawings of embryos which we will discuss later - see Appendix B Unsubstantiated claims by evolutionists)

I recently went to hear Dr. Guliuzza, M.D. speak. He made the point that the proponents of evolution make a concerted effort to show everything as "simple." Why? The reason is simple as well: if everything is simple, then it is easy to see how evolution could occur. It is when you discover how complex life is and how finely tuned the cosmos is that you begin to question how this could ever have come about by random, undirected chance. It could not and did not.

DNA - The Instructions

(deoxyribonucleic acid)

DNA is much more complex than our illustration will convey in this discussion. You can see from the illustration that the four nucleotide units are adenine, thymine, cytosine, and guanine, or as we will refer to them from now on, A, T, C, and G. The sides known as the phosphate backbone of the helix consist of alternating deoxyribose sugars and phosphates. Inside the DNA are the instructions for all the proteins necessary to create a complete living being. We now know that there is a functional system above DNA that directs its expression in specific parts of the body, known as epigenetics. We will discuss this phenomenon later.

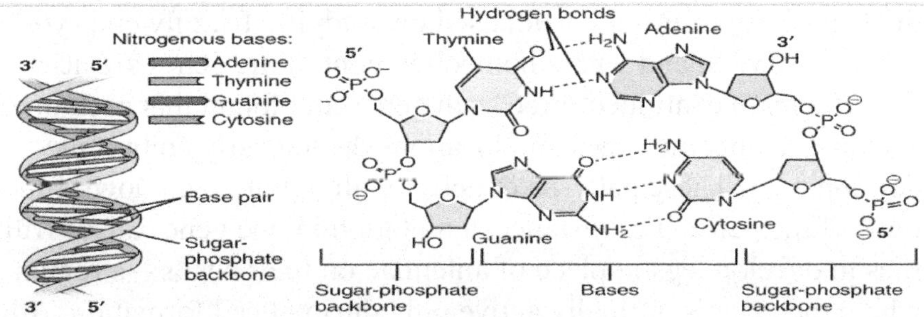

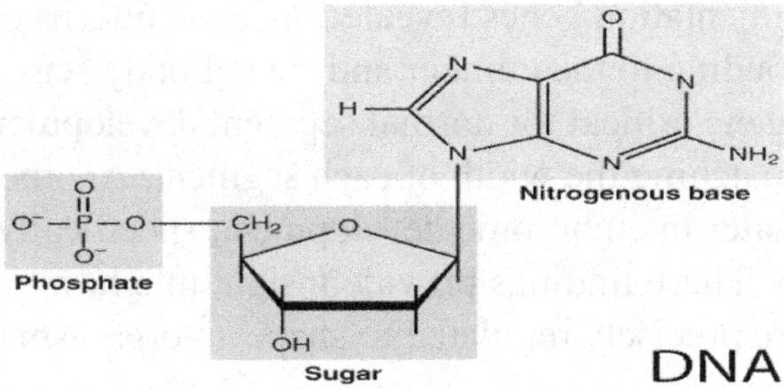

DNA Nucleotides

The base pairs always match up Adenine with Thymine and Guanine with Cytosine. The genes have all the genetic possibilities for a given entity. There are two expressions of the living entity: the genotype and the phenotype. The distinction is that the genotype is the individual's genetic code, and the phenotype is the physical body or expression of the code, such as eye color, skin color, hair color, etc. During embryo development, the phenotype is determined. The "body type" (an entity's symmetry, segmentation, and limbs) is determined at that time. There are several theories on how the body type is determined. While working with fruit flies, scientists have discovered what they call the "homeobox," which may explain part of this phenomenon. Additionally, we now know that epigenetics is responsible for the expression of these genes in making specific body parts. We will discuss how it works later on in this chapter.

A recently discovered segment of DNA known as the homeobox is believed to play a key role in directing the development of the body plan in a wide range of animals. Research led by Walter Gehring at the University of Basel has focused on studying fruit fly embryos to observe when and where homeobox-containing genes are active. His team examined three such genes and found that each one activates at specific times and locations during early embryonic development. These patterns of activity align with the known functions of the genes. For instance, a mutation in one gene causes fruit flies to develop legs in place of antennae on their heads—a region where the gene is normally active only during head formation. Additional studies on segmentation genes revealed striped patterns of gene activity, corresponding to the number and size of body segments. In one case, a gene critical for normal segment development was active in stripes matching the width of each segment. Another gene showed activity later in embryonic development, specifically in the posterior region. These findings provide insight into how genetic instructions are precisely regulated to shape an organism's body plan.

There is so much, which is not known, that it is difficult to write a book and be current when new discoveries are happening all the time. We now find that there may be two locations that affect the body plan of a developing embryo, that body plan established in a growing embryo may come from something other than the genetic material in the nucleus.

Dr. Ray Bohlin, in The Natural Limits to Biological Change, highlights a significant revelation in developmental biology: much of the information guiding early embryonic development is not located in the nucleus, but rather in the cytoplasm and cell membrane. While this was suspected a decade prior, it has since been strongly confirmed. This discovery challenges the long-held belief that the nucleus—and specifically genetic material like DNA—is the sole source of developmental information. Bohlin explains that there is at least one additional source of biological instruction at work, implying that successful changes to an organism's body plan would require coordinated modifications in both genetic and non-genetic systems. Because these two sources may be unrelated and function independently, achieving such coordinated changes through natural processes becomes exceedingly difficult. This insight underscores the limitations of evolutionary explanations that rely solely on genetic mutation and reinforces the view that the complexity of life involves multiple layers of information—each requiring precision and foresight.

There are an amazing number of traits that are turned off in the genes. What that means is that you have the genes for all possible combinations of phenotype expression. If you are Caucasian, you express fair skin in the phenotype (physical body traits). However, you have in your genome (DNA) everything needed for dark skin. It happens that the expression of that gene is turned off. We are all of the species called "Homo sapiens"; one skin color is not better than another. This would be a big surprise to Hitler and others like him who claim superiority based on skin, hair, or eye color. You might find viewing the motion picture "The Mortal Storm," made in the early 40s before the U.S. joined World War II, very enlightening. The movie shows how faulty science can have dramatically bad effects on people's thinking.

In the 1920s version of a science book used in the Scopes trial in 1925 (A Civic Biology by George Hunter) the book stated that:

> If we follow the early history of man upon the earth, we find that at first he must have been a little better than one of the lower animals.
>
> At the present time there exists upon the earth, five races or varieties of man. **the highest type of all, the Caucasians, represented by the civilized white inhabitants of Europe and America**. [my emphasis]

It is no wonder why we had racism in this country and why Hitler could justify the mass murder in Germany of Jews, blacks, and anyone not perfectly healthy. In the early 20th century, based on the faulty logic of evolutionary thinking, Ota Benga, an African pygmy, was put on display as an example of evolution and placed in a cage like the monkeys.

> Ota Benga (circa 1883[1] – March 20, 1916) was a Congolese man, a Mbuti pygmy known for being featured in an anthropology exhibit at the Louisiana Purchase Exposition in St. Louis, Missouri in 1904, and in a controversial human zoo exhibit in 1906 at the Bronx Zoo. Benga had been purchased from African slave traders by the explorer Samuel Phillips Verner, a businessman hunting Africans for the Exposition. He traveled with Verner to the United States. At the Bronx Zoo, Benga had free run of the grounds before and after he was "exhibited" in the zoo's Monkey House. Except for a brief visit with Verner to Africa after the close of the St. Louis Fair, Benga lived in the United States, mostly in Virginia, for the rest of his life. - Ota Benga. (2015, June 28). In Wikipedia, The Free Encyclopedia. Retrieved 23:31, June 28, 2015, from Https://en.wikipedia.org/w/index.php?title=Ota_Benga&oldid=669051835

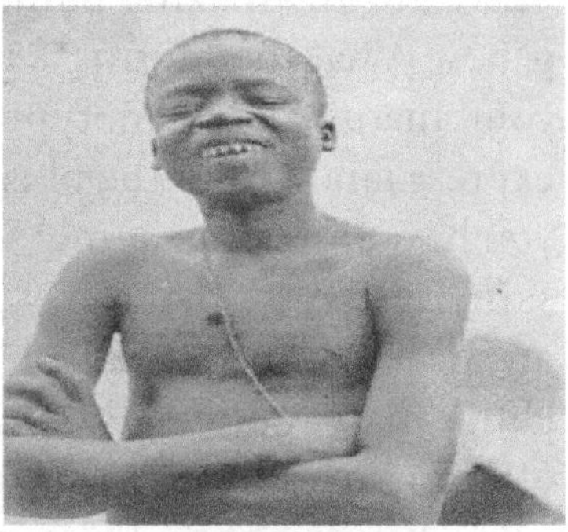

The Darwinists have been looking for decades for missing links in the fossil record. Apart from a few misrepresented examples and those proven to be fraudulent, there are no transitional fossils. (See Appendix B: Unsubstantiated Claims by Evolutionists.) Because of the limits of the gene pool, you never see one species transform into another. Many experiments have been done to try to disprove that fact. The Drosophila, or fruit fly, as it is commonly known, has been studied and mutated for many decades, and yet it remains a fruit fly. There have been many variations, but in the end, it remains a fruit fly. To this date, there have never been instances where one species type becomes another. Darwin's "***On the Origin of Species***" remains unsolved. It appears that God spoke correctly in Genesis.

> *Genesis 1:24 "And God said, Let the earth bring forth the living creature after his kind, cattle, and creeping thing, and beast of the earth after his kind: and it was so".*

Created kinds are seen as having common descent if they are reproductively compatible. Pierre-Paul Grassé a French zoologist puts it succinctly: In ***Evolution of Living Organisms***, Pierre-Paul Grassé argues that mutations observed in bacteria and viruses do not lead to genuine evolutionary advancement, but rather represent hereditary fluctuations—minor shifts to the right or left around a stable genetic baseline. These variations, he contends, produce no lasting evolutionary change. To illustrate this, he compares cockroaches—one of the oldest surviving insect groups—with the fruit fly Drosophila. Despite having undergone countless mutations over time, cockroaches have remained virtually unchanged since the Permian period of the Paleozoic Era. This stability, despite the potential for genetic variation, suggests that mutation alone is insufficient to drive the kind of large-scale evolutionary transformation proposed by Darwinian theory.

We see micro-evolutionary changes in a species every day; just look at the variety of dogs. The genome (DNA) allows for great variety in phenotype. Consider the variety of beaks in Darwin's Galapagos finches. God, in His infinite wisdom and design, has built a code so complex that it can produce an almost limitless variety of phenotypic expressions. However, offspring are always within a type and never outside that type. The mule, for instance, is a cross between a horse and a donkey; however, the mule is sterile, so there would be no chance of offspring and, therefore, the end of that potential evolutionary change.

> The Modern Synthesis is a remarkable achievement. However, starting in the 1970s, many biologists began questioning its adequacy in explaining evolution. Genetics might be adequate for explaining micro-evolution, but micro-evolutionary changes in gene frequency were not seen as able to turn a reptile into a mammal or to convert a fish into an amphibian. Micro-evolution looks at adaptations that concern only the survival of the fittest, not the arrival of the fittest. Scott Gilbert, John Opitz, and Rudolf Raff, "*Resynthesizing Evolutionary and Developmental Biology*", Developmental Biology, 173, Article no. 0032, 1996, p. 361

A relatively new field of study is Baraminology. Baraminology is the classification and study of kinds, founded in 1990 by Kurt Wise. Baraminology is based on the presupposition that God created many kinds of organisms, as described in the Biblical book of Genesis. Our body has about 75 to 100 trillion cells. There are 46 chromosomes: 23 from mom and 23 from dad. Recently, we have seen many people claiming that they are female or male and are undergoing "sex change" operations to make it a reality. One male (who was a female) claimed to be pregnant. This, of course, would be a physical impossibility. The male body does not have the equipment or the genetics to accomplish that task. We will explore this in great detail in Volume II on reproduction. It was eventually revealed that she was claiming to be male. Females have two X chromosomes and males have an X and a Y chromosome. Since your DNA is found in every one of the trillions of cells, no matter what the surgeons do, the DNA of that individual will always remain either XX in females or XY in males, making the idea of there being any other gender purely psychological and not physical. Wishful thinking does not change reality. There are about ten billion base pairs in the chromosomes. Your entire blueprint is packed in these DNA molecules. The DNA is found in every one of those 100 trillion cells! DNA is so insanely complex and so tightly packed that it is awe inspiring. Werner Gitt gives us an idea of just how small the DNA molecule is:

> "After a number of intelligent scientists had worked for many years developing ever better microfilm they fit the entire Bible on one 32 X 33 mm film. Amazing! However, that same space covered with DNA would hold information equivalent to 7.7 million Bibles"! Werner Gitt, *In the Beginning Was Information*, 1997, p. 192-194

The amino acid is a three-letter word known as a codon (see Table of amino acids under Appendix A). These three-letter words define which amino acid is used in building a specific protein. You will remember from the preface that we spoke about what happens if the coding for phenylalanine has just one defective base pair. If the base pair (one pair out of three pairs that define the amino acid) at position 408 in the chain is coded as CG, you have a normal protein and a normal person (assuming there are no other errors). If the base pair located in that same position in the chain is coded as TA, then you have Phenylketonuria (PKU) (the inability to metabolize phenylalanine) and thus the propensity towards mental retardation.

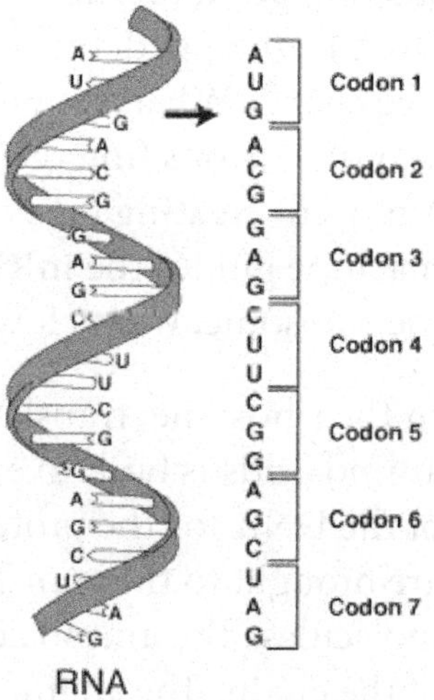

RNA

Aspartame, an artificial sweetener sold under the trade names "Equal" and "NutraSweet," is a non-food source of the amino acid phenylalanine. When consumed, aspartame is metabolized by the body into several byproducts, one of which is phenylalanine. For individuals with phenylketonuria (PKU)—a genetic disorder that impairs the body's ability to break down phenylalanine—this can pose a health concern. Although the buildup of phenylalanine from aspartame is typically less severe than that from dietary proteins, it can still contribute to elevated levels in people with PKU, necessitating caution in their consumption of products containing this sweetener.

Transcription

The flow chart on the next page is an oversimplified illustration of the transcription and translation process. We know that DNA contains the instructions for building every cell in your body. How does a cell know when it needs to create a protein? Why does a cell need to die? What process causes a cell to divide and create a new cell?

The overly simplified version of translation is this: Something (the operating system) decides that a specific protein is needed. So the process of transcription and translation begins. Something (the operating system) tells the process what protein is needed, and so the process directs the RNA polymerase to go to a specified starting point on the DNA string. Using helicase enzymes, the DNA is unzipped, breaking hydrogen bonds between nucleotides (see chart). When termination is reached, the RNA is processed, and the introns (intergenic regions), non-coding sections, are removed, creating mRNA (messenger RNA). Another molecular machine guides the mRNA through the nuclear pore complex to the ribosome, where it will be translated.

In the ribosome, translation of the mRNA takes place. A string of amino acids is built to exact specification from the instructions of the DNA for the protein that is needed. These amino acids are brought to the translation process by tRNA (transfer RNA) molecules. The amino acid AUG or methionine acts as the start of the chain. The chain will end when the process reaches a UGA, UAG, or UAA codon, known as stop codons; these do not correspond to any amino acid, and the chain stops. The tRNA will place the exact right amino acid on the chain. The next step is to fold the string of amino acids into a protein. This protein must be folded to exact specifications, or the protein is not usable, or it can be fatal.

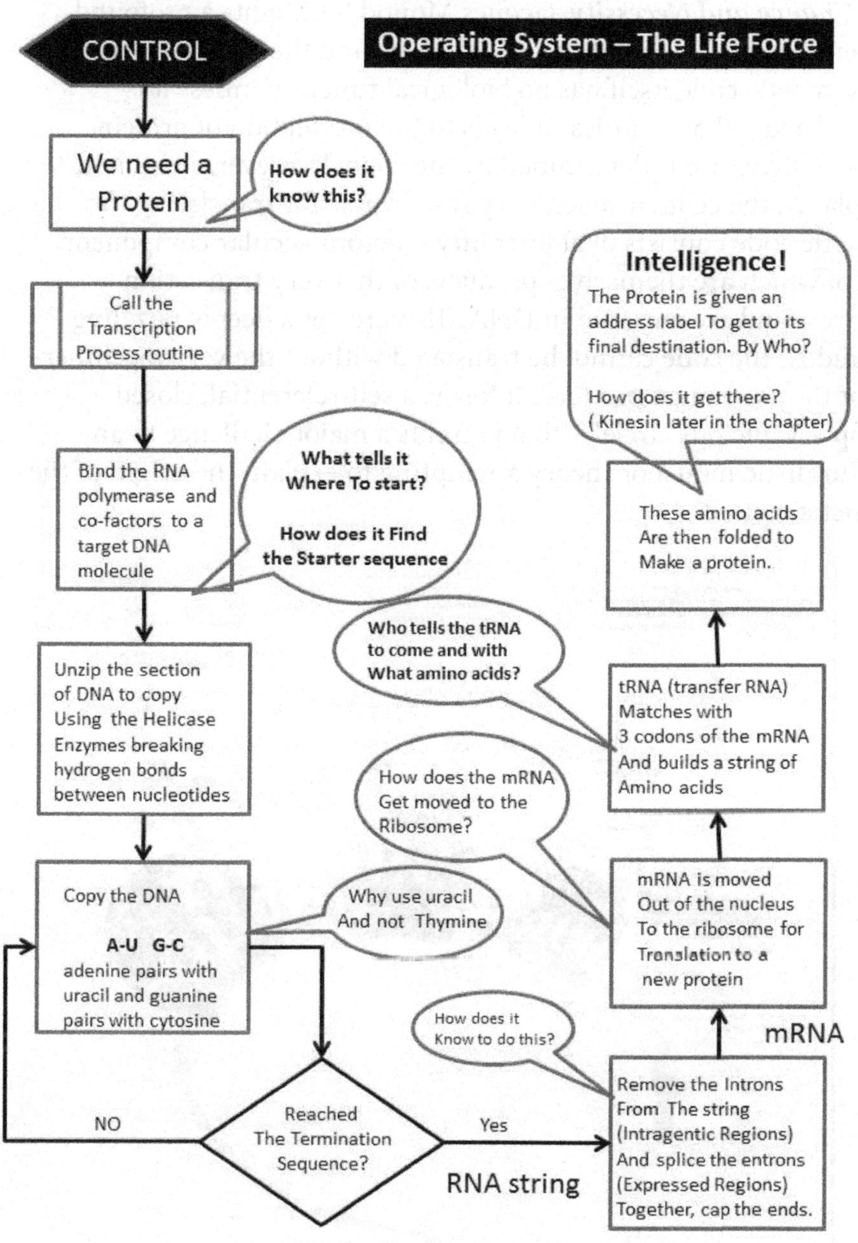

The diagram on last page is an illustration of how the mRNA string is translated into a protein.

In ***Chance and Necessity***, Jacques Monod highlights a profound dilemma at the heart of the origin of life and the genetic code. The genetic code itself has no biological function unless it is translated—that is, unless it leads to the production of proteins whose structure is determined by the code. However, as Monod explains, the cellular machinery responsible for translating the genetic code consists of at least fifty macromolecular components, all of which are themselves products of that very translation process and are encoded in DNA. This creates a deeply puzzling paradox: the code cannot be translated without the very machinery that the code must produce. It forms a self-referential, closed loop—a "vicious circle"—that presents a major challenge to any naturalistic model or theory attempting to explain the origin of the genetic code.

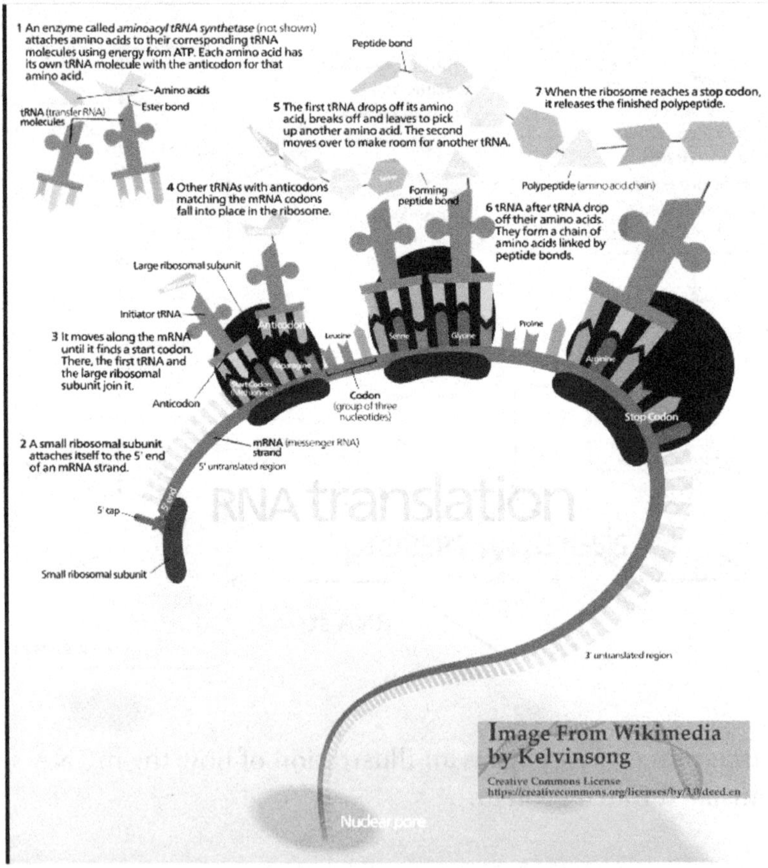

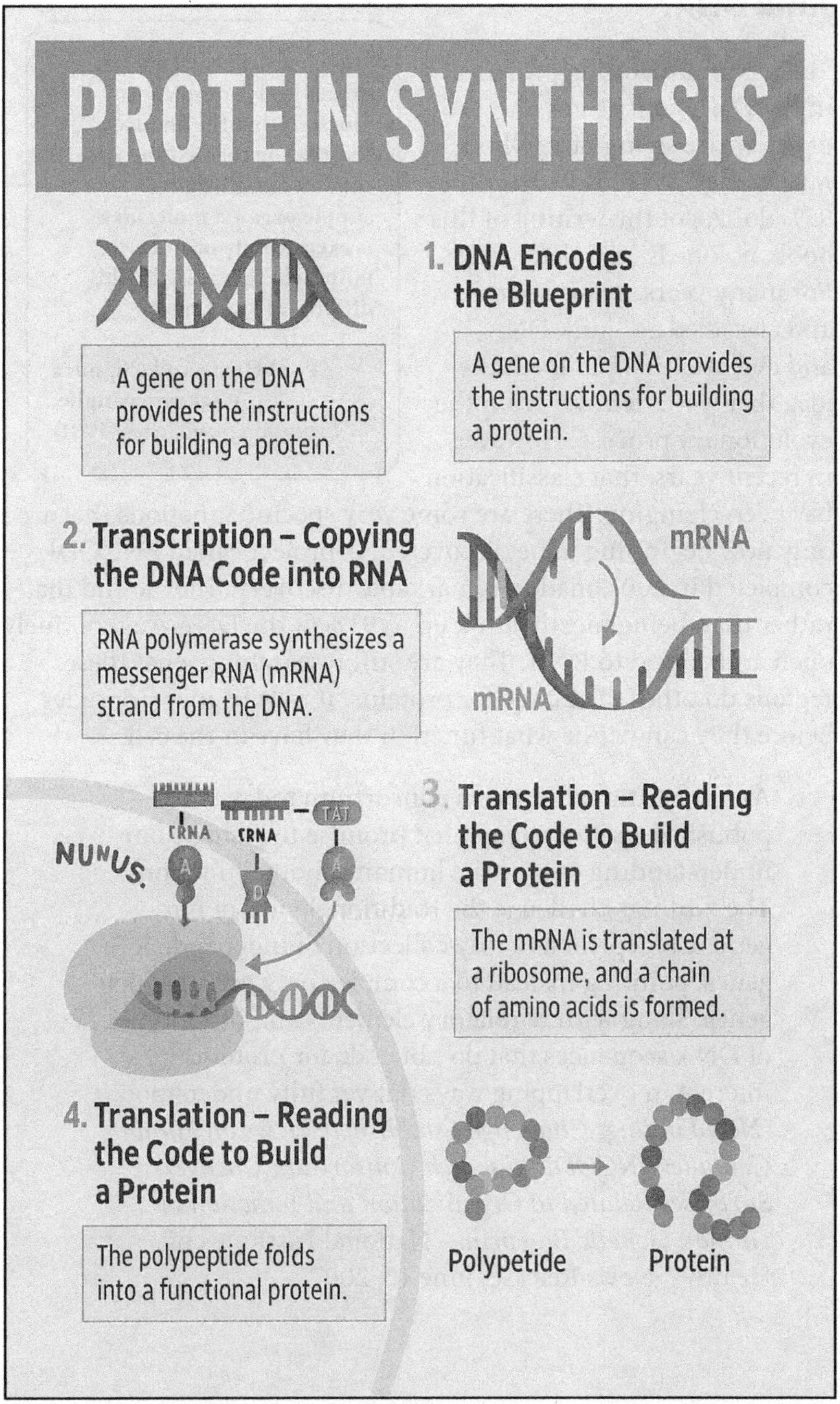
PROTEIN SYNTHESIS
1. DNA Encodes the Blueprint
A gene on the DNA provides the instructions for building a protein.
A gene on the DNA provides the instructions for building a protein.
2. Transcription – Copying the DNA Code into RNA
RNA polymerase synthesizes a messenger RNA (mRNA) strand from the DNA.
mRNA
mRNA
3. Translation – Reading the Code to Build a Protein
The mRNA is translated at a ribosome, and a chain of amino acids is formed.
NUNUS
TAT
4. Translation – Reading the Code to Build a Protein
The polypeptide folds into a functional protein.
Polypetide
Protein

Junk DNA?

> **"Biochemical systems are exceedingly complex, so much so that the chance of their being formed through random shufflings of simple organic molecules is exceedingly minute, to a point where it is insensibly different from zero."**
>
> **Fred Hoyle and Chandra Wickramasinghe,** *Evolution from Space* **(1981)**

The region for coding proteins in the DNA constitutes only 2% of the entire string of 3 billion nucleotides. What does the other 98% do? As of the writing of this book, no one is completely sure. For many years, it was rejected and classified as "junk DNA," and evolutionists put forth the idea that it was leftover from the evolutionary process. However, in recent years, that classification has been changing. There are some very specific functions that are only now beginning to be discovered. A project called ENCODE, completed in 2007, made a remarkable discovery. They found that rather than being mostly junk, up to 93% of the DNA was routinely open and copied to RNA. They are still researching what these regions do other than code for proteins. It may be many decades before they can define what function they have in the cell.

> An international research consortium today published a set of papers that promise to reshape our understanding of how the human genome functions. The findings challenge the traditional view of our genetic blueprint as a tidy collection of independent genes, pointing instead to a complex network in which genes, along with regulatory elements and other types of DNA sequences that do not code for proteins, interact in overlapping ways not yet fully understood. *-New Findings Challenge Established Views on Human Genome: ENCODE Research Consortium Uncovers Surprises Related to Organization and Function of Human Genetic Blueprint* - National Institutes of Health - News Release, June 13, 2007

It has been postulated that the operation of the human body may be like an operating system. Certainly, we know that coding for proteins is not necessarily contiguous in the DNA. So it is like the operating system on your computer, which finds a file on your hard disk. The file itself may not be contiguous for a variety of reasons, but the operating system knows how to find all of its parts. When coding for a protein, something in the cell knows how to construct the messenger RNA for translation into a protein. The question here is: what directs the "building" of a human from conception to birth? What directs the order of building the separate parts? How do the cells know where to go or what to be? This may be somewhat unknown, but with the discovery of epigenetics, we know that when a heart cell is made, all other cell types are permanently turned off in that cell. The DNA remains the same in all of the trillions of cells in your body, except for the methyl groups and histones affected by epigenetics. The methylation of parts typically acts to suppress the transcription of a section of DNA, making it effectively unavailable to manufacture a specific protein.

> A single gene can potentially code for tens of thousands of different proteins... It's the way in which genes are switched on and off, though, that has turned out to be really mind-boggling, with layer after layer of complexity emerging. (~ Le Page, "*Genome at 10,*" New Scientist, 6/ 16/ 10).

> During the early debates about the Human Genome Project, researchers had predicted that only a few percent of the human genome sequence encoded proteins, the workhorses of the cell, and the rest was junk. We now know that this conclusion was wrong. - Eric D. Green, director of NHGRI (National Human Genome Institute) from an article in Science Daily

To complicate matters further, the coding of a protein can overlap with other codes on the same genetic string space. When you are conceived, a set of events begins to take place that is a true miracle. The sperm and the egg join, and you are given 23 pairs of genes that will determine your phenotype. The DNA "parts list," as I have heard Dr. Jonathan Wells put it, is used in the process of building a human being. Let's pause for a minute and consider the probability that life arose by chance. We can excuse Darwin to some degree because he did not have the advantage of knowing what we do today about the extreme complexity of the cell. On the other hand, scientists today have no excuse. They do have knowledge of the complexity of life. Some will honestly admit they do not understand how life arose; others do not admit it. What they cannot explain, they either dismiss or fabricate some elaborate story. Most times, they come up with something completely bizarre, like, for instance, "punctuated equilibrium" (see the section in the introduction called Do You Believe in Magic). It is not a matter of evidence; the evidence is clear; it is how they choose to interpret the information. According to Hoyle and Wickramasinghe, "life cannot have had a random beginning." In Evolution from Space, Sir Fred Hoyle and N. Chandra Wickramasinghe confront the improbability of life arising through purely random processes. They point out that the formation of even the basic biological components—such as the approximately 2,000 enzymes required for life—through random molecular trials is astronomically unlikely. They calculate the chance of assembling all these enzymes by chance as one in 10^40,000, a probability so minuscule that it defies plausibility, even if the entire universe were composed of organic "soup." Given this, they argue that any theory with a probability even slightly greater than this should be considered more credible. They propose that the idea of life being assembled by an intelligent designer carries a vastly higher probability and should be seen as the most reasonable explanation for the origin of life. In their view, the resistance to this conclusion is not due to scientific evidence, but rather to psychological and philosophical biases against the concept of intelligent design.

It has been established that anything greater than 1060 is a statistical impossibility. How can the evolutionist have such faith in a theory that is so statistically small? I think Hoyle is correct; it is psychological rather than scientific. Yet, we are told by people like Eugenie Scott, Executive Director of the National Center for Science Education (at the time of this research), that evolution is science and creation is not. It would take as much of a miracle for life to have arisen by chance as for an all-powerful God. Their humanistic (atheistic) belief is a religion whose god is blind chance.

I learned about chance at a very early age. I went to the Iowa State Fair when I was 12. I wanted to win a watch I saw at one of the carnival booths. The guy assured me I could win. He even showed me how to throw the ring over the bottles. I tried and failed, but he kept encouraging me to try again. I tried over and over but was never able to win. When I was completely out of money, I quit in shame and defeat. All the money I had worked so hard for and saved was gone. I spent the next three days at the fair with no money. I have not gambled since that time. I do not believe in chance with the odds at 1:1,040,000. I had a better chance of winning the watch, even though the system was rigged so I could not win

> "Mathematicians agree that any requisite number beyond $1{:}^{1060}$ has, statistically, a zero probability of occurrence (and even that gives it the benefit of the doubt!). Any species known to us, including the smallest single-cell bacteria, have enormously larger numbers of nucleotides than 100 or 1000. In fact, single cell bacteria display about three million nucleotides, aligned in a very specific sequence. This means, that there is no mathematical probability whatever for any known species to have been the product of a random occurrence—random mutations (to use the evolutionist's favorite expression)." L. Cohen, *Darwin was Wrong* (1984), p. 205.

At this point, we have talked about the wonders of DNA and how its three billion nucleotides could not have arisen by chance, as well as some of the ways DNA is used in the cell to create proteins. The question we should ask is: how does the cell get the energy to produce activity? We will address the energy mechanism in the next chapter. In The Nature and Origin of the Biological World, E. Ambrose highlights a fundamental paradox in molecular biology: enzymes, which are essential for the functions of life, operate only because their specific protein chains are coded by DNA. Yet DNA itself cannot replicate without the assistance of protein enzymes. This creates a classic "chicken and egg" dilemma—both DNA and proteins are interdependent, and neither can function or originate in isolation from the other. This mutual dependence raises profound questions about how such a system could have arisen through gradual, unguided processes, pointing instead to the necessity of an already functioning system from the very beginning.

Mitochondrial Eve

In most of our cells, there are several mitochondrial organelles; a liver cell has approximately 2,000, while a red blood cell has none. This organelle is responsible for the production of energy in the cell, known as adenosine triphosphate, or ATP. The mitochondrion is unique in that it has its own independent genome (DNA). This is commonly referred to as mitochondrial DNA (mtDNA). This mtDNA, unlike DNA found in the nucleolus of the cell, remains mostly unchanged from generation to generation.

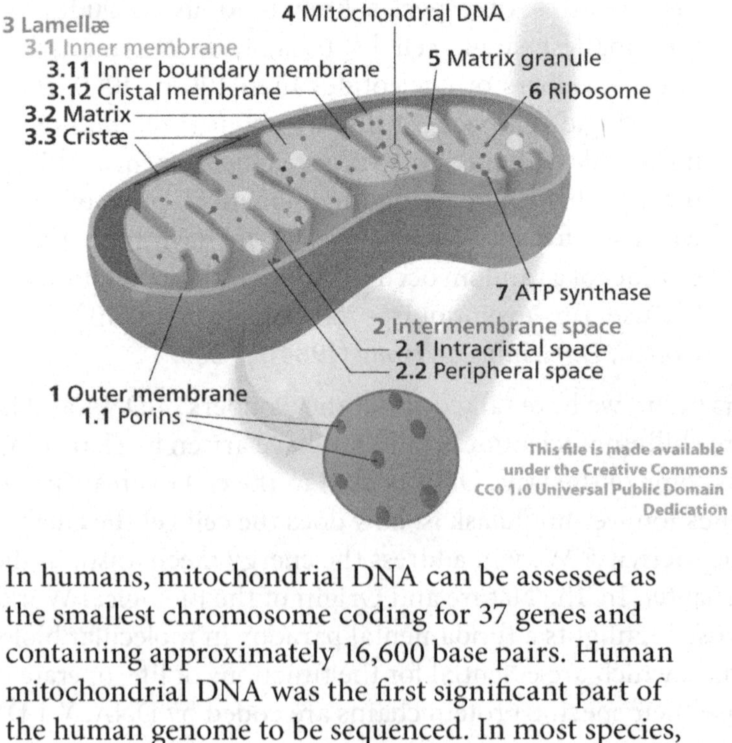

In humans, mitochondrial DNA can be assessed as the smallest chromosome coding for 37 genes and containing approximately 16,600 base pairs. Human mitochondrial DNA was the first significant part of the human genome to be sequenced. In most species, including humans, mtDNA is inherited solely from the mother. "*Mitochondrial DNA: The Eve Gene*". Bradshaw Foundation. Bradshaw Foundation. Retrieved 5 November 2012.

Dr. Bryan Sykes wrote a book called The Seven Daughters of Eve. Dr. Sykes investigated the mtDNA found in a 5,000-year-old frozen Italian specimen. He found that the DNA was the same as that in modern Europeans. In one case, he found that the mtDNA was the same, making them and the others genetically linked.

The title of Sykes's book comes from the fact that 95% of Europe's population comes from just seven haplogroups. Haplogroups define differences in human mitochondrial DNA. Sykes identified them by the letters U, X, H, V, T, K, and J, and further gave them names: Ursula, Xenia, Helena, Velda, Tara, Katrine, and Jasmine. The Europeans could now trace their lineage back to a common mother.

In 1997, a paper entitled A High Observed Substitution Rate in the Human Mitochondrial DNA Control Region, by Thomas J. Parsons, was published in Nature Genetics. The authors compared the mtDNA of many mother-child pairs and found that mutations in mtDNA occur about 20 times more rapidly than previously thought. Based on these measurements, they calculated that Mitochondrial Eve lived only about 6,500 years ago.

- The Berkeley team measured differences in human mitochondrial DNA.
- Parsons's study measured the rate of change in human mitochondrial DNA.

> When these two sets of empirical data are put together, they yield a date for Mitochondrial Eve of about 6,500 years. This is about the time the Bible gives for when the real Eve lived. The only real reason this figure is rejected is because it agrees with the Bible, while being contrary to evolution.- "Human migration." CreationWiki, The encyclopedia of creation science.

Amazing DNA Doctors

An article by Kelly Hollowell, J.D. Ph.D. titled 'Mutations Mean Death and Disease,' states that before a cell can divide, it must replicate the 46 chromosomes, making an exact copy. If there is one single error in 150,000,000 nucleotides, problems may arise, such as sickle cell anemia among others. She explains that there are repair enzymes, referred to as 'DNA doctors,' that scan the DNA spiral for errors in the code. If they find an error, they replace it with corrected DNA code. If the repair enzymes do not function, the outcome is typically cancer. In fact, literally thousands of human diseases are associated with genetic mutations; there are currently 4,500 different diseases linked to genetic mutations. Other diseases implicated by mutations in the DNA sequence include hereditary high blood levels of cholesterol and cystic fibrosis. From what we can observe, mutations are rarely, if ever, beneficial. Darwin's theory that small mutations can account for all the variety and complexity of life is highly improbable, making it unreasonable. Now we understand that epigenetics plays a very large part in disease as well.

> The enzyme encircles the double helix to repair a broken strand of DNA. DNA ligase is responsible for repairing the millions of DNA breaks generated during the normal course of a cell's life. Without molecules that can mend such breaks, cells can malfunction, die, or become cancerous. DNA ligases catalyse the crucial step of joining breaks in duplex DNA during DNA repair, replication and recombination. Wikipedia

The question is, how can such a complex and exact process be generated by random, undirected chance? First, if it could be generated by chance, the system would never survive long enough to be perfected. Errors in the mutated code would be so mangled that they would either cause massive damage or death. Secondly, where did the intelligence come from that could recognize errors in the code? Third, what intelligence could lead to the function of self-healing without a designer? These functions are so critical that there isn't a way they could come about by small incremental steps. The giant leaps of punctuated equilibrium do not explain it any better. It has never been observed, so the idea is purely a matter of immense faith in an outdated philosophy. I am not convinced that chance could ever generate a system so finely tuned and so complex. You never see a natural process where a system left on its own can replicate itself and correct any errors that may occur in its function.

The problem with evolutionary thinking is that it cannot explain how mutations and natural selection account for the synergistic nature of living systems. The key word here is "systems." First, for natural selection to have any value, the mutation must be advantageous, in place, and completely operational. Second, it would require massive mutations in various parts of DNA, since these must affect several functions at once to be of any use. This is an example of what Michael Behe calls "irreducible complexity," something that evolutionists would like to deny exists. There is an interrelationship between functions in living systems. Take the human eye; if it were to be developed in small incremental steps as evolutionists would have us believe, it would have to be joined by the simultaneous development of many other systems that support it.

To have functional sight requires (this is an extreme oversimplification):

<u>A way to convert light</u> into electrical impulses, and the pathway to communicate to the brain.

<u>Muscles to control</u> the pupil's size as the intensity of light changes. The brain function must also be fully operational to direct the muscles to reduce the pupil, or the eye would be damaged by intense light.

<u>Eyelids to protect the eye</u>, the brain function to tell the eye when to close or open, and the path to communicate. Not having the ability to close the eyelid will result in blindness; being closed all the time does not allow clear vision, making the function virtually useless.

Lacrimal glands, the source of tears, to keep the eye lubricated, clean away debris, bring oxygen to the cornea, provide proteins to coat the eye, the brain function to control it, and the pathway to communicate with the glands.

Brain function, which converts electrical and chemical pulses to something understandable (this may be the most difficult of all). This visual information is also stored in the brain (temporarily or long term).

Without all of these systems working at once, you have nothing that will allow sight. The system is there, all at once, or the so-called fortuitous mutations would be worthless. We have already described that, from what we can currently observe, there are virtually no mutations that improve a system. Most mutations are neutral at best, harmful, or lethal in the worst case. Also, the DNA would have to be changed in the reproductive cells to be passed on to the next generation. The entire thing is a fantasy. After God created all things, the natural process took over to give us variety, which was built into the genome from the beginning (micro-evolution).

Epigenetics

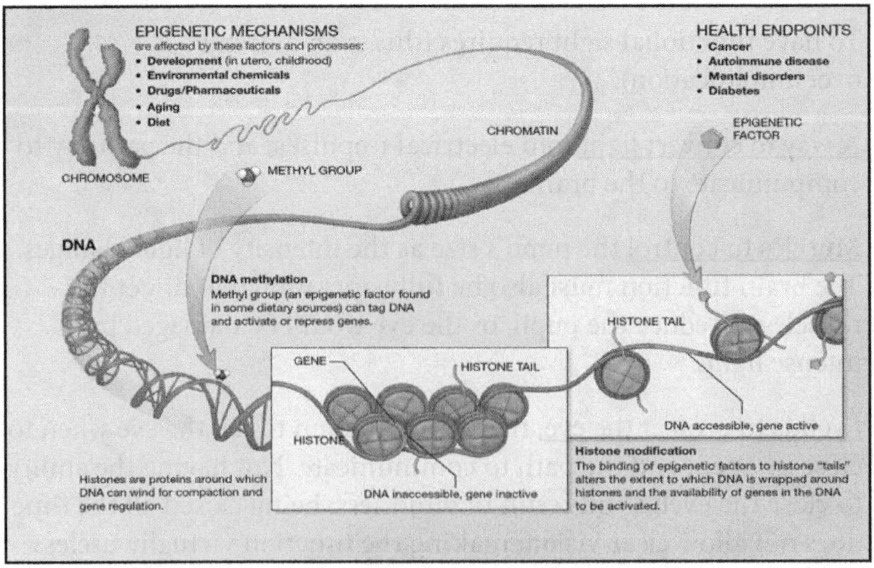

> A single gene can potentially code for tens of thousands of different proteins... It's the way in which genes are switched on and off, though, that has turned out to be mind-boggling, with layer after layer of complexity emerging. - Michael Le Page, "*Genome at 10: Information overload*" New Scientist Magazine.

Epigenetics is a flourishing field of biological research that deepens the complex mechanisms through which the genes are regulated, offering critical insights on how genetic instructions are expressed. Defined in general, the epigenetics refers to the study of hereditary changes in the gene activity that do not involve alterations of the DNA sequence below. This term embodies a range of biochemical processes that influence the gene expression, often compared to a set of molecular "switch" that can modulate the level of gene activity. These Dimmer switches can improve or suppress the brightness of genetic instructions, ultimately determining the measure in which specific behaviors or biological processes are manifested.

Understanding of epigenetics is particularly significant for teenagers, a development phase characterized by profound biological and psychosocial transitions. During adolescence, individuals experience a myriad of changes while moving towards physical maturity. These changes are not dictated exclusively by genetic information codified in DNA; Rather, the expression of specific genes during this transformative period can be significantly influenced by epigenetic mechanisms. For example, the expression of the genes responsible for emotional regulation, the development of the brain and hormonal balance can be developed by external factors, including diet, stress and social interactions. Therefore, while an individual's DNA provides the essential project for development, the epigenetic factors regulate how this project is emanated vividly or thinly.

The implications of epigenetics extend beyond simple academic curiosity; They have practical relevance for understanding health and well -being throughout life. For example, research has shown that environmental exposures during critical development windows, such as in uterus or in childhood, can induce epigenetic changes that influence the susceptibility to various diseases further on in life. Factors such as maternal nutrition, exposure to toxins and the psychosocial environment can lead to lasting epigenetic changes that influence the models of gene expression during the life of an individual. For teenagers, awareness of these influences can favor a proactive position towards life choices, underlining that not only genetic predispositions but also environmental factors play a vital role in modeling their future.

In essence, while the DNA acts as a set of basic instructions, the epigenetics provides a picture through which these instructions are dynamically regulated. This regulation improves our understanding of the biological diversity and the individuality of the traits, putting the interconnection between genetic inheritance and environmental conditions bare. It underlines the importance for teenagers - which are often in a crucial moment to form their identities - to recognize the agency that hold in modeling their genetic expression through informed lifestyle choices. This realization can encourage greater improvement for the complex interaction between genetics and the environment, ultimately guiding teenagers towards decisions more attentive to health while navigating in this fundamental period of their lives. In summary, the study of epigenetics not only illuminates the mechanisms underlying gene regulation, but also authorizes individuals to understand their unique biological narratives in the wider context of environmental influences. The gene expression is the fundamental process through which the information codified in our DNA is translated into functional proteins that perform essential roles in the body. To understand this complex mechanism, it is useful to use a simple analogy of a cookbook. In this analogy, DNA represents the entire cookbook, containing all the recipes (genes) necessary to create various dishes (proteins). Not all recipes are always necessary, just like not all the genes of our DNA are active in each cell.

The gene expression occurs in two main phases: transcription and translation. The transcription is similar to the opening of the cookbook and the copy of a specific recipe on a separate piece of paper. In biological terms, this process takes place in the nucleus of the cell, in which the DNA filament unfolds and a messenger molecule called RNA Messenger (MRNA) is summarized based on the sequence of the gene. This mRNA acts as a temporary and portable copy of the recipe, since the original DNA remains safely protected within the nucleus.

Once the transcription is completed, the MRNA comes out of the nucleus and enters the cytoplasm of the cell, where the second stage - translation occurs. During the translation, the cell reads the mRNA and uses ribosomes (chefs in our cooking analogy) to assemble amino acids in a specific order, creating a protein according to the MRNA recipe. This protein can therefore perform various vital functions for the cell and the body, such as the construction of tissues, the catalization of chemical reactions or the regulation of cell processes.

Epigenetics introduces a further layer of complexity to gene expression. While the DNA sequence remains unchanged, the accessibility and activity of some genes can be influenced by epigenetic modifications. Two key mechanisms of the epigenetic regulation are the methylation of the DNA and the modification of the histone.

DNA methylation provides for the addition of a methyl group (chemical tag) to specific regions of the DNA molecule, often leading to the silence of genes. Imagine this how to place a adhesive note on a recipe in the cooking book that says "do not use". This change can prevent the transcription mechanism from accessing the DNA, thus closing the expression of that gene.

The modification of the histone refers to the changes made to the proteins (histones) around which DNA is enveloped. These changes can alter the structure of the chromosome, making it more strictly or freely packaged, effectively controlling the accessibility of the associated genes. Continuing with our example of cookbooks, the modification of the histone can be compared to reorganizing the shelves of a kitchen to make it easier to reach some recipes or hide them, influencing which recipes the chef decides to cook.

According to Kanherkar et al. (2014), gene expression is not a static process; Rather, it is a dynamic interaction between genetic predispositions and epigenetic changes influenced by various factors during life. From prenatal development to old age, environmental elements such as nutrition, stress and exposure to toxins can induce epigenetic changes that influence gene expression. This alignment of genetic and epigenetic factors underlines the nuanced nature of development and health, illustrating that our biological destiny is not dictated exclusively by our genetic code but is significantly modeled by the environment in which we exist. Epigenetics plays a crucial role in prenatal development, where the development fetus is highly sensitive to environmental influences. The investigation indicates that the epigenome, which regulates gene expression without altering the underlying DNA sequence, can be significantly affected by the factors in the surroundings of a mother, particularly those experienced during pregnancy. It has been shown that maternal nutrition, stress levels and exposure to toxins induce epigenetic modifications that can affect not only the fetus but also the individual's health throughout his life.

Maternal nutrition is a key factor that influences epigenetic results. The study by Indrio et al. (2017) emphasizes the impact of early nutritional environments on long -term health, revealing that the availability of certain nutrients can modify the epigenetic marks that regulate the patterns of gene expression. For example, folate, an essential vitamin B, plays a crucial role in DNA methylation processes, which are a form of epigenetic modification. The adequate maternal intake of folate has been associated with optimal fetal development and a reduced risk of congenital abnormalities. On the contrary, a vital nutrient deficiency during critical development periods could lead to adverse epigenetic changes that could predispose the child to health problems later in life, such as obesity, diabetes or cardiovascular disease.

In addition to nutrition, maternal stress is another significant factor that can cause epigenetic changes in the development fetus. High levels of stress hormones, such as cortisol, can permeate the placental barrier, affecting the development of fetal brain and general growth. Research suggests that chronic stress during pregnancy can lead to alterations in gene expression related to the response to stress and emotional regulation, which may have lasting implications for psychological and physiological health of offspring. Studies have illustrated that children exposed to high levels of stress in the uterus have a greater risk of developing anxiety disorders and behavioral problems, underlining the deep impact of the prenatal environment on mental and physical well -being.

It has been shown that exposure to environmental toxins, such as heavy metals, pollutants and chemicals for endocrine disruptors, has harmful epigenetic effects during pregnancy. For example, maternal exposure to high levels of lead has been correlated with changes in DNA methylation patterns in offspring, which can lead to deteriorated cognitive function and behavioral problems. These findings highlight the importance of fighting for a prenatal environment -free environment, since epigenetic modifications induced by harmful substances can affect the long -term health trajectory of the development fetus.

The interaction of maternal nutrition, stress and exposure to toxin illustrates the complexity of epigenetic influences during prenatal development. These environmental factors can alter the expression of genes that are critical for growth and development, fundamentally configuring the health and well -being of the individual throughout their life course. Consequently, it is essential to establish a healthy prenatal environment. Interventions aimed at improving maternal nutrition, managing stress and minimizing exposure to harmful substances can positively influence fetal epigenoma, resulting in better health results for future generations. Understanding epigenetics not only underlines the importance of prenatal health, but also provides valuable information about possible strategies to improve maternal and child health throughout useful life. Critical development windows during childhood and adolescence are fundamental periods in which epigenetic mechanisms can generate lasting effects on health and behavior. These phases exhibit high plasticity, whereby the genome remains receptive to external stimuli, which allows environmental factors to play an important role in the configuration of the epigenetic landscape of an individual. Simpkin et al. (2016) emphasize that the experiences found during early development are not simply transitory events; On the other hand, they can lead to modifications in the patterns of gene expression that persist throughout the life of an individual.

The concept of "sensitive periods" or "critical periods" refers to specific deadlines during which the influence of environmental factors, from nutritional status to psychosocial stressors, has a pronounced impact on gene regulation. For example, Mychasiuk and Metz (2016) elaborate how the stress experienced during these early stages can facilitate epigenetic changes that affect the hypothalamic-political-adrenal (HPA) axis, a key component involved in the response to stress. Such alterations in the HPA axis can predispose people to mental health problems, including anxiety and depression, highlighting the lasting nature of early life experiences.

In addition, adverse environmental factors, such as exposure to toxins, maternal nutrition or socioeconomic state, can lead to epigenetic modifications that alter development and neuronal behavior. Simpkin et al. (2016) has shown that poor nutrition during critical development periods can lead to alterations in DNA methylation patterns, which significantly affects brain development and other organ systems. This evidence underlines the notion that what happens in the training years can have deep implications for health trajectories in later life.

In addition to negative influences, positive experiences during critical windows can similarly induce beneficial epigenetic changes. Qualitative environments, characterized by positive interactions for parents and children, safe attachments and enriched learning conditions, have been related to favorable epigenetic modifications. These experiences encourage resilience and can promote healthier development results as individuals make the transition from childhood to adolescence and beyond. Mychasiuk and Metz (2016) provide convincing evidence that early enrichment can pave the way to improve neuronal connectivity and functioning, which could have cascade effects on cognitive skills and emotional regulation.

It is crucial to understand that epigenetics does not work in isolation; Rather, it incorporates the dynamic interaction between genes and environmental influences. This reality implies that interventions aimed at improving the first experiences of children could lead to deep epigenetic changes, potentially reverse negative trajectories initiated by adverse conditions. In essence, promoting healthy environments during childhood and adolescence is not simply a matter of biological predisposition, but a convincing call to action for political leaders, educators and caregivers equally.

As the field of epigenetics continues to expand, more and more reveals the importance of considering the integral interaction of genetics and the environment in the critical stages of development. By understanding these dynamics, more emphasis can be placed on strategies to promote positive epigenetic results for future generations, shaping health and behavior so that they extend much beyond childhood and early childhood. Environmental influences play a crucial role in the formation of epigenetic modifications which can have a significant impact on the expression of genes throughout the life of an individual. Understanding these influences requires an exploration of various factors, including stress, food and family relationships, which collectively contribute to the complex interaction between our genes and in the environment. Goodrich et al. (2016) provide a complete study illustrating how exposure to a range of environmental conditions during critical development windows, as before birth and throughout adolescence, modifies epigenetic profiles and affects global development.

The prenatal period is particularly sensitive to environmental factors. Maternal stress has been shown, for example, has deep effects on fetal development. Goodrich et al. (2016) discuss how maternal stress leads to changes in fetal epigenome, influencing genes that regulate stress and neurodevelopment responses. Glucocorticoid stress hormone, cortisol, can modify DNA methylation models in the fetus, thus modifying the gene expression linked to the regulation and development of stress. These alterations have not only an impact on the immediate development of the child, but can also predispose them to various psychological and physiological disorders later in life.

Food factors are another important environmental influence. Nutritional choices during pregnancy can considerably influence epigenetic mechanisms. Goodrich et al. (2016) highlight studies demonstrating that the maternal intake of certain nutrients, such as Folate and omega-3 fatty acids, can lead to favorable epigenetic changes which promote healthy neurode development in offspring. Conversely, deficiencies or poor food choices can trigger unfavorable epigenetic changes, perhaps resulting in developmental disorders and long-term health implications. The concept of "nutritional epigenetics" underlines the importance of food not only on traits and results for health, but also as a modulator of the expression of genes through epigenetic paths.

Family relationships and the quality of social environments during critical development stages, such as childhood in adolescence, can influence epigenetic changes through mechanisms such as social stress and attachment. Goodrich et al. (2016) emphasize that the negative family dynamics, including exposure to chronic conflicts or negligence, can change gene expression profiles linked to emotional regulation and the response to stress. The evidence suggests that the unfavorable experiences of childhood (ACE) can lead to persistent changes in DNA methylation schemes, thus influencing the risk of developing mental health problems and behaviors unsuitable for adulthood.

Peer and social support interactions can play a protective role against unwanted results, promoting positive epigenetic changes. Positive social experiences can lead to a decrease in the expression of stress -related genes, while negative experiences can exacerbate stress responses through epigenetic mechanisms. Thus, the social environment, shaped by family relationships and interactions between peers, can serve as a significant modulator of the expression of genes, influencing the development trajectories of the beginning of life in adolescence.

The work of Goodrich et al. (2016) illustrates the deep impact of environmental factors on epigenetic processes. The complex interaction between stress, diet and family relationships affects not only the expression profiles of immediate genes, but also has implications for long -term health results. Understanding these mechanisms is essential to promote healthier environments that promote positive development results thanks to targeted interventions that consider the multifaceted nature of human development. Epigenetic changes play a significant role in health results, particularly during adolescence, a crucial period of development characterized by rapid physiological and psychological changes. Understanding the interaction between genetic predispositions and environmental factors is essential to understanding how conditions such as obesity and mental health disorders manifest themselves. Recent longitudinal studies, especially those conducted by Wahlqvist et al. (2015) and Suarez et al. (2018), illuminated these complex relationships, emphasizing the importance of environmental influences on gene expression.

Wahlqvist et al. (2015) investigated the association between epigenetic modifications and obesity in adolescents, demonstrating that environmental factors, including eating habits, physical activity and socioeconomic status, are an integral part of gene expression linked to metabolic processes. Through their longitudinal approach, the researchers followed a teen cohort over several years, analyzing their epigenetic markers alongside lifestyle variables. The results indicated that adolescents exposed to diets rich in sugars and fats exhibited distinct epigenetic modifications in genes related to fat storage and metabolism. These changes were still exacerbated by a sedentary lifestyle, emphasizing the need to consider environmental factors in the context of genetic predisposition in relation to obesity.

When evaluating blood samples at different times, Wahlqvist et al. It has revealed that adolescent years serve as a critical window, where epigenetic changes can predispose individuals to obesity in later adulthood. Notably, these epigenetic markers have proved to be reversible, providing information on lifestyle interventions to positively influence gene expression-related gene expression. This paper emphasizes the significance of preventive health measures in adolescence aimed at environmental and genetic factors.

Similarly, the study by Suarez et al. (2018) expanded the understanding of how epigenetic mechanisms contribute to mental health disorders during adolescence, focusing on depression and anxiety. This longitudinal study examined a diverse population of adolescents, some of which suffered childish adversities such as neglect or emotional abuse. The findings indicated that adolescents exhibited distinct epigenetic modifications in genes associated with the response to stress and neuroplasticity. The changes in gene expression were closely linked to adverse experiences, highlighting the importance of the first environmental factors in the formation of mental health results.

Suarez et al. It also elucidated that the moment of environmental exhibitions plays a crucial role in epigenetic programming. Adolescents who experienced stress early in their lives were more susceptible to the development of mental health problems, as epigenetic changes influenced the essentials of neurological development essential for emotional regulation. In addition, the study incorporated interventions, suggesting that support environments and positive practices for parents could mitigate these adverse epigenetic changes, promoting resilience and improving the results of mental health.

These studies collectively illustrate that a comprehensive understanding of adolescent health should take into account the intricate relationship between genetics and environmental influences. Epigenetic modifications serve as a crucial bond, revealing how external factors can modify gene expression, subsequently impacting health results such as obesity and mental health disorders. Therefore, neglecting the role of environmental factors will lead to an incomplete understanding of the etiology of these conditions during adolescence. The ideas obtained from research for epigenetics not only advance in our understanding of the complexity of human health, but also emphasize the critical importance of creating support environments that can positively influence gene expression and, finally, health care trajectories. By synthesizing the discussions presented throughout this research article, it is evident that epigenetics serves as a crucial intersection between genetic predispositions and environmental influences. By understanding epigenetics, adolescents can appreciate as genes, although fundamental to biological identity, are not unchanging health or behavior determinants. Instead, gene expression is dynamically regulated by epigenetic mechanisms, such as DNA methylation and histones modification, which can be significantly influenced by environmental factors found before birth and throughout life.

The implications of this research are deep. He emphasizes the notion that lifestyle choices, eating habits, stress levels and social interactions can contribute to the epigenetic changes that affect not only the health and good -of the individual, but also that of future generations. For example, exposure to toxins, nutritional deficiencies or psychosocial stressors during critical periods of development, such as in the womb, may lead to lasting changes in gene expression, potentially predisposing individuals to various health problems later in life. This relationship highlights an inherent flexibility in our biology that, instead of making us the US passive recipients of our genetic destination, enables us to play an active role in the formation of our health through informed choices.

In addition, continuous exploration of epigenetic mechanisms provides fertile ground for future research directions. Scientists are increasingly recognizing that epigenetic changes can contribute to complex diseases such as cancer, cardiovascular disease and mental health disorders. This understanding opens avenues for innovative therapies that can achieve epigenetic changes associated with these conditions, offering hope of more personalized and effective treatment strategies.

The dynamic nature of epigenetic regulation raises intriguing questions about human development throughout life. As researchers continue to investigate how lifestyle and environmental variables interact with genetic information, it will be crucial for young people to stay informed and actively engage in their health and well -being. Knowledge about epigenetics not only enriches the understanding of biology, but improves awareness of personal potential to positively influence one's genetic expression.

In short, the relevance of epigenetic studies continues to grow as we seek to unravel the complexities of human health and development. By filling genetic factors with environmental influences, epigenetics elucidate the mechanisms by which our life choices can sculpt our biology. Thus, involvement with this field of study encourages adolescents to reflect on their environments and decisions, promoting a proactive approach to shaping their future health results.

All the instructions for building and maintaining the human body are present. So, when a protein is needed, the software finds the information it needs to build the protein and then transcribes it into a string. This will be translated later by combining amino acids into a string, which is then folded into a protein. Epigenetics is the software that can change access to that data so that it is effectively "turned off." The original DNA is not changed; only the access to it is. All of the trillions of cells in your body have the same DNA; it is the access to that data that changes what happens in any cell type.

For example, when a child is forming in the mother's uterus, it is critical that the software can differentiate between developing heart muscle cells and nerve cells. To accomplish that (we are still wondering how it knows to do that), it must turn off parts of the DNA that code for heart tissue when developing the brain and its neuron cells. It is clear that there is intelligence that goes far beyond undirected random chance. There are over 200 different cell types in the human body, which means that there are more than 200 different "programs" differentiating the various cell types. The methyl markers effectively turn off all the genes that are not necessary for making that cell.

Epigenetics affects the phenotype (the physical expression—eye color, skin color, etc.), while not affecting the genotype (DNA). Epigenetic research has shown that environmental factors have an effect on epigenetic expression. The old saying that "you are what you eat" may be truer than we have thought. Diet and exercise do play a part in how genes are switched on and off.

Additionally, our parents can also pass on parts of their epigenetics to offspring. Is this what we are being told in Exodus 34?

> *Keeping mercy for thousands, forgiving iniquity and transgression and sin, and that will by no means clear the guilty; visiting the iniquity of the fathers upon the children, and upon the children's children, unto the third and to the fourth generation. Exodus 34:7 (KJV)*

To illustrate the role of epigenetics, children born to mothers who were in the early stages of pregnancy during the Dutch famine from 1944-1945 had children who were obese and had heart issues. Painter R.C. Roseboom T.J. Bleker O.P. *Prenatal exposure to the Dutch famine and disease in later life: an overview. Reproductive Toxicology*

Epigenetics and Cancer

It is now believed that epigenetics could be the cause of some cancers. Our DNA has a tumor suppression gene; if that gene is methylated (turned off), then cancer is most likely the result of uncontrolled cell growth. Another possibility is that the genes responsible for DNA repair get turned off; this too would cause DNA damage and result in uncontrolled growth, given the right circumstances. This research is in its infancy compared to DNA research. Who knows what wonders we will find in the future as God reveals more of His handiwork to us.

> In each cell, there are multiple operating systems, multiple programming languages, encoding/ decoding hardware and software, specialized communications systems, error detection/ correction systems, specialized input/ output for organelle control and feedback, and a variety of specialized "devices" to accomplish the tasks of life. - Dr. Donald E. Johnson - quoted from Stephen Williams. *What Your Atheist Professor Doesn't Know (But Should)*

Citations:

Simpkin, Andrew J. Laura D. Howe, Kate Tilling, Tom R. Gaunt, Oliver Lyttleton, Wendy L. McArdle, Susan M. Ring, Steve Horvath, George Davey Smith, and Caroline L. Relton. The epigenetic clock and physical development during childhood and adolescence: longitudinal analysis from a UK birth cohort. International journal of epidemiology 46, no. 2 (2017): 549-558. Https://academic.oup.com/ije/article-abstract/46/2/549/2907775

Simpkin, Andrew J. Gibran Hemani, Matthew Suderman, Tom R. Gaunt, Oliver Lyttleton, Wendy L. Mcardle, Susan M. Ring et al. Prenatal and early life influences on epigenetic age in children: a study of mother–offspring pairs from two cohort studies. Human molecular genetics 25, no. 1 (2016): 191-201. Https://academic.oup.com/hmg/article-abstract/25/1/191/2384645

Keverne, Eric B. Donald W. Pfaff, and Inna Tabansky. Epigenetic changes in the developing brain: Effects on behavior. Proceedings of the National Academy of

Sciences 112, no. 22 (2015): 6789-6795. Https://www.pnas.org/doi/abs/10.1073/pnas.1501482112

Indrio, Flavia, Silvia Martini, Ruggiero Francavilla, Luigi Corvaglia, Fernanda Cristofori, Salvatore Andrea Mastrolia, Josef Neu et al. Epigenetic matters: the link between early nutrition, microbiome, and long-term health development. Frontiers in pediatrics 5 (2017): 178. Https://www.frontiersin.org/articles/10.3389/fped.2017.00178/full

Mychasiuk, Richelle, and Gerlinde AS Metz. Epigenetic and gene expression changes in the adolescent brain: What have we learned from animal models?. Neuroscience & Biobehavioral Reviews 70 (2016): 189-197. Https://www.sciencedirect.com/science/article/pii/S0149763416301373

Kanherkar, Riya R. Naina Bhatia-Dey, and Antonei B. Csoka. Epigenetics across the human lifespan. Frontiers in cell and developmental biology 2 (2014): 49. Https://www.frontiersin.org/articles/10.3389/fcell.2014.00049/full

Lin, Xinyi, Ives Yubin Lim, Yonghui Wu, Ai Ling Teh, Li Chen, Izzuddin M. Aris, Shu E. Soh et al. Developmental pathways to adiposity begin before birth and are influenced by genotype, prenatal environment and epigenome. BMC medicine 15 (2017): 1-18. Https://link.springer.com/article/10.1186/s12916-017-0800-1

Goodrich, Jaclyn M. Dana C. Dolinoy, Brisa N. Sánchez, Zhenzhen Zhang, John D. Meeker, Adriana Mercado-Garcia, Maritsa Solano-González, Howard Hu, Martha M. Téllez-Rojo, and Karen E. Peterson. Adolescent epigenetic profiles and environmental exposures from early life through peri-adolescence. Environmental epigenetics 2, no. 3 (2016): dvw018. Https://academic.oup.com/eep/article-abstract/2/3/dvw018/2415066

Wahlqvist, Mark L. Stephen A. Krawetz, Nico S. Rizzo, Maria Gloria Dominguez-Bello, Linda M. Szymanski, Shari Barkin, Ann Yatkine et al. Early-life influences on obesity: from preconception to adolescence. Annals of the New York Academy of Sciences 1347, no. 1 (2015): 1-28. Https://nyaspubs.onlinelibrary.wiley.com/doi/abs/10.1111/nyas.12778

Mulligan, Connie J. Early environments, stress, and the epigenetics of human health. Annual Review of Anthropology 45, no. 1 (2016): 233-249. Https://www.annualreviews.org/content/journals/10.1146/annurev-anthro-102215-095954

Bozack, Anne K. Sheryl L. Rifas-Shiman, Diane R. Gold, Zachary M. Laubach, Wei Perng, Marie-France Hivert, and Andres Cardenas. DNA methylation age at birth and childhood: performance of epigenetic clocks and characteristics associated with epigenetic age acceleration in the Project Viva cohort. Clinical Epigenetics 15, no. 1 (2023): 62. Https://link.springer.com/article/10.1186/s13148-023-01480-2

Suarez, Anna, Jari Lahti, Darina Czamara, Marius Lahti-Pulkkinen, Polina Girchenko, Sture Andersson, Timo E. Strandberg et al. The epigenetic clock and pubertal, neuroendocrine, psychiatric, and cognitive outcomes in adolescents. Clinical epigenetics 10 (2018): 1-12. Https://link.springer.com/article/10.1186/s13148-018-0528-6

DNA Mutations

The following are just a few mutations that occur in human DNA. There are many mutations that occur in humans; none of them are helpful. If you are healthy, thank the Lord for His great design in creation.

Cystic fibrosis - caused by the presence of mutations in both copies of the gene for the protein cystic fibrosis transmembrane conductance regulator (CFTR) - O'Sullivan, BP; Freedman, SD (30 May 2009). "Cystic fibrosis."

The CFTR protein is a channel protein that controls the flow of H2O (water) and Cl (chlorine) - ions into and out of cells inside the lungs. When the CFTR protein is working correctly, as shown in Panel 1, ions freely flow in and out of the cells. However, when the CFTR protein is malfunctioning as in Panel 2, these ions cannot flow out of the cell due to a blocked channel. This causes Cystic Fibrosis, characterized by the buildup of thick mucus in the lungs.

The mutation in the CFTR protein is caused when the three nucleotides of the amino acid phenylalanine at the 508th position on the protein are deleted. The CFTR gene is found on chromosome 7 and is 1,480 amino acids long. When I was writing this, there was no cure known for the mutation.

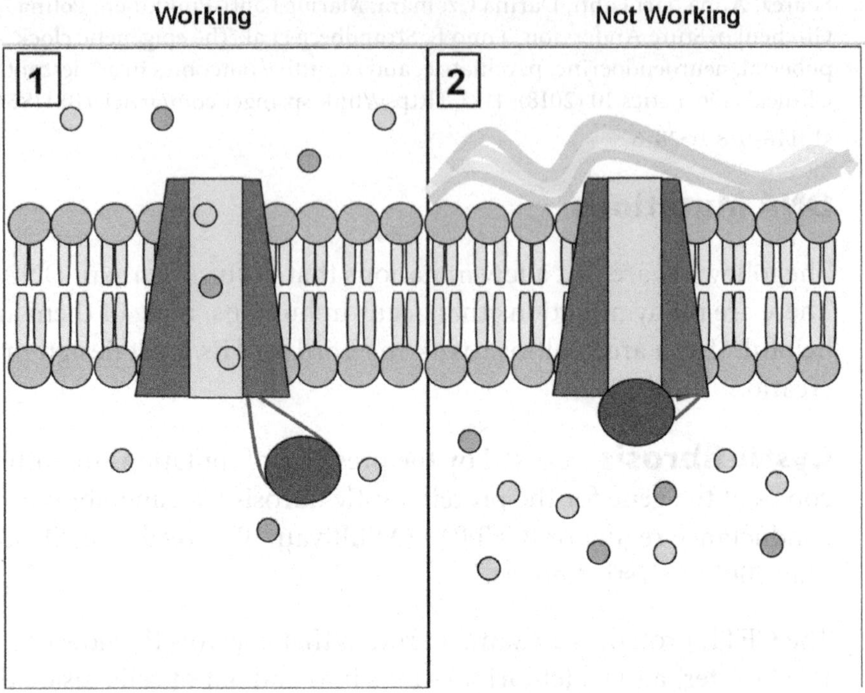

Sickle cell anemia

Hemoglobin is a protein in red blood cells that carries oxygen throughout the body. Sickle cell disease is a disorder of the red blood cells in which the cells take on a sickle-like shape, thus giving the condition its name. The mutation occurs when a defective gene is inherited from both parents. If only one mutation is passed on, the person is considered only a carrier of the gene. This mutation is the result of a single nucleotide change, where glutamic acid is substituted by valine at position 6 of the β-globin gene (from a GAG to GTG codon on the coding strand).

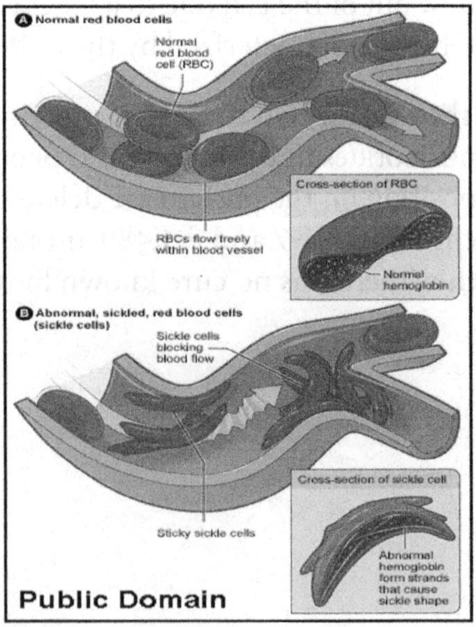

The cells in your body require a steady flow of oxygen, and the hemoglobin in red blood cells carries this oxygen. Normal red blood cells are flexible and can easily move through both large and small blood vessels. The sickle cell shape is not flexible and can cause blockages, preventing oxygen from reaching the cells.

Phenylketonuria

Phenylalanine is necessary for growth in infants and for nitrogen equilibrium in adults. Normally, the production of the protein consists of 451 amino acids in its chain, with arginine at the 408th position. If the 408th position is miscoded, it can result in phenylketonuria (PKU) or the potential for mental retardation. You can see from the illustration below that if just one base pair is coded incorrectly, it can lead to a potential disaster! This illustrates how very precise God is in His creation; otherwise, most of us would be mentally retarded.

	Normal	PKU
DNA at codon 408	x x x **C** G G x x x	x x x **T** G G x x x
	x x x **G** C C x x x	x x x **A** C C x x x
Amino Acid at 408	arginine	tryptophan

Downs Syndrome: also known as trisomy 21

This malady comes from a mutation when a third copy of gene 21 appears in the DNA, yielding 47 chromosomes. This causes confusion during the manufacture of proteins, resulting in either too many proteins being made or too few. Sometimes, chromosomes are incorrectly distributed into the

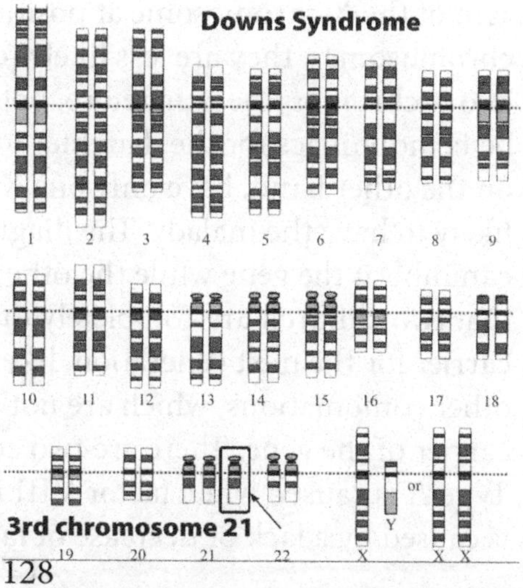

egg or sperm cells during meiosis. When this happens, one cell may get two copies of a chromosome, while another cell gets none. The incorrect distribution of chromosomes is called nondisjunction. - Http://learn.genetics.utah.edu/content/chromosomes/diagnose/"

Hemophilia:

This problem is caused when the F8 gene (coagulation factor VIII, procoagulant component) on the X chromosome has been mutated. This factor is manufactured in the liver and is found circulating in the bloodstream. When damage occurs in a blood vessel, this factor and others set off chain reactions that form a blood clot. When an injury occurs, if the F8 gene is not operating correctly, blood clotting does not form properly, causing excessive bleeding. F8 is located on the long arm of the X chromosome at position 28. Since women have two X chromosomes, they are less likely to have this issue. Females have two X chromosomes; therefore, both the mother and father must be hemophiliacs for the daughter to inherit the problem. Males, on the other hand, have only one X chromosome, so they are more likely to have the malady. The diagram above shows how one son can inherit the gene while the other children do not. It also shows that two children are completely unaffected and one becomes a carrier for the next generation, like her mother. There are many other combinations, which are not shown here, if the father is a carrier of the gene. There are two common types of hemophilia. Type A is caused when factor VIII is lacking or decreased. Type B is caused by a lack or decrease of factor IX (Antihemophilic Factor

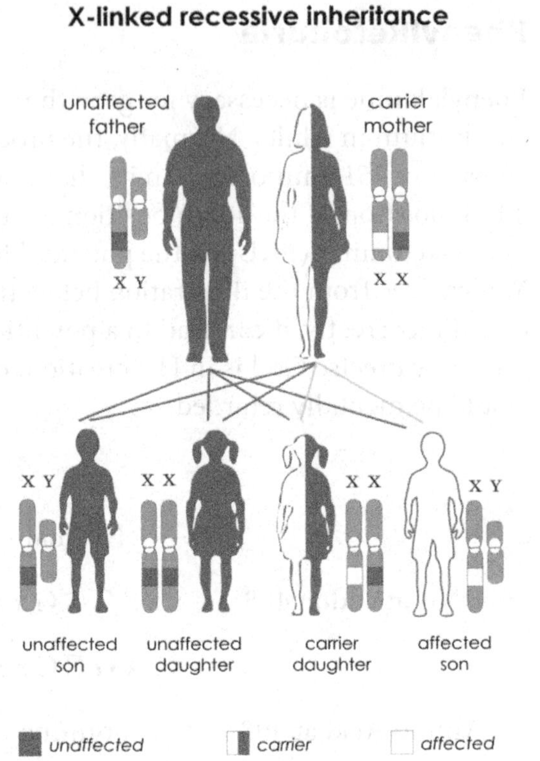

B or Christmas Factor). DNA mutations are virtually never helpful. God designed us from the beginning. It is clear that we are fearfully and wonderfully made, and it is mutations that distort, not create better humans! How would evolution occur when mutations normally cause the body to operate incorrectly? The design is clear to anyone not biased by years of propaganda. To believe that serendipitous accidents could create such a machine as the human body is untenable to all but the willingly ignorant. Hemophilia was known as "the royal disease." Queen Victoria (late 19th century) was a carrier. Her son, Prince Leopold, was a hemophiliac. Most children during that period died early. However, because Leopold was protected all the time, he lived to maturity. Leopold did finally marry. His daughter was a carrier but not a hemophiliac. His two sons did not have it, but the third generation had one hemophiliac. Queen Victoria was believed to have the relatively rare Hemophilia B."

Albinism:

Albinism is the complete lack of pigment in the skin, eyes, and hair. This is caused by the absence or defect of the enzyme tyrosinase, which is used in the production of melanin that gives the skin its color. This malady is also associated with vision problems.

PGK1 gene:

This gene makes an enzyme called phosphoglycerate kinase. This enzyme is critical in glycolysis, which is the process of breaking down glucose for the creation of energy in the form of ATP. The molecular location of this protein is on the X chromosome.

The disorder known as phosphoglycerate kinase deficiency affects the body's ability to break down glucose. There are two basic forms of the problem: the hemolytic form and the myopathic form. It is a rare inherited metabolic disorder.

> The condition results from mutations in Pgk1, the gene encoding PGK1, and twenty mutations have been identified. - Beutler E (Jan 2007). "*PGK deficiency*". *British Journal of Haematology*

The hemolytic form is characterized by red blood cells being destroyed prematurely, known as hemolysis. Hemolysis (from the Greek αἷμα (hemo) meaning "blood" and λύσις (lysis) meaning "a loosening"). The consequences can range from mild to critical. The symptoms are fatigue, shortness of breath, and jaundice; in critical cases, it can lead to a stroke or seizures. The myopathic form affects the muscles. This causes weakness or cramping during exercise.

> During exercise, muscle tissue can be broken down, releasing a protein called myoglobin. This protein is processed by the kidneys and released in the urine (myoglobinuria). If untreated, myoglobinuria can lead to kidney failure. - *Genetics Home Reference* - Http://ghr.nlm.nih.gov/condition/phosphoglycerate-kinase-deficiency

Again, I think you can see just how fragile we are as humans. Mutations do not add information, nor do they make things better.

It is clear that we were designed from the beginning. As small mutations occur in each generation, we are not getting better from generation to generation; we are getting worse.

The ultimate symbiotic relationship argument exists in the fact that we humans are male and female. There is no way to incrementally, step by step, evolve the combination in two separate entities that have such a strict symbiotic relationship. It cannot get any simpler than that. Why or how could chance mutations create two entities that work so well together and evolve them separately?

Evolution is based on blind chance. It has no information about male or female; it has to be accomplished by small incremental mistakes, backed up by selection. Females have all of the amazing features that are needed to grow and protect a new developing human. All the parts must be there, in working order, or there will be no more humans. The male has all the parts and chemistry needed to create a living being in combination with the female. If evolution were true, and a single non-gender entity was capable of self-generation, why would there be any need to evolve a male and a female gender? The idea of evolution is ludicrous!

Additionally, the human male is the only gender capable of creating both male and female since the male has both X and Y chromosomes, while the female only has X. That in itself proves that woman came from man in the beginning, just as Genesis tells us.

Tim

Gen 2:22
And the rib, which the LORD God had taken from man, made he a woman, and brought her unto the man.

Mark 10:6
But from the beginning of the creation God made them male and female.

THE NOT SO SIMPLE CELL

4

The Not So Simple Cell

An adult human body contains tens of trillions of cells. Because trillions of these cells die every day, the body always has a repair kit on hand to make duplicate copies as rapidly as the old cells disappear.

Richard Swenson, *More Than Meets the Eye: Fascinating Glimpses of God's Power and Design (LifeChange)*

In This chapter:

Cell Membrane and Organelles
Cell types
Energy Generation - ATP
Electrical Energy - Sodium/Potassium Pumps
Programmed death of a cell - Apoptosis
Cell Transport Systems
Stem Cells

"I grew up believing in this (evolution) Myth and I have felt—I still feel—its almost perfect grandeur. Let no one say we are an unimaginative age: neither the Greeks nor the Norsemen ever invented a better story. But the Myth asks me to believe that reason is simply the unforeseen and unintended by-product of a mindless process at one stage of its endless and aimless becoming. The content of the Myth thus knocks from under me the only ground on which I could possibly believe the Myth to be true. If my own mind is a product of the irrational, how shall I trust my mind when it tells me about evolution?"

C.S. Lewis, *Christian Reflections*, Wm. B. Eerdmans Pub. Co. p. 89, 1975.

In Darwin's time, according to Ernst Haeckel, the cell was nothing more than a "simple little lump of an albuminous combination of carbon" and a mere "homogeneous globule of plasm." How far can one be from the true nature of the cell? To be fair, Darwin and Haeckel had no way of knowing the true nature of the cell; they can be excused. However, today's scientists are without excuse. Today, if we were able to see the number of complex molecular machines operating in the cell continuously, as we see any machine, like a transparent airplane, for instance, I think people would quickly change their thinking about life. But because we cannot see it or feel it, the mystery remains, and Darwinists can continue to cling to their idea of naturalism. I believe that, in the future, someone like Bones in Star Trek will exclaim how belief in evolution was such a barbaric idea.

In his essay *The Cell as a Collection of Protein Machines*, Bruce Alberts explains why large protein assemblies within cells are often referred to as "protein machines." He draws a parallel between these biological structures and human-engineered machines, noting that both consist of highly coordinated moving parts designed to perform specific functions efficiently. In these cellular assemblies, the interactions between molecules are not random but follow a controlled sequence of events—reaction C depends on reaction B, which in turn depends on reaction A. This ordered, interdependent behavior mirrors the way machines operate in the macroscopic world, where each component must function in harmony with others to achieve a desired outcome. Alberts emphasizes that this mechanistic precision within cells underscores the remarkable design and complexity of biological systems.

Cell or Plasma Membrane

The cell membrane protects the cell by determining what is allowed to come in and what is sent out. Oxygen, nutrients, and signals come in; waste, carbon dioxide, and signals go out. The membrane is made up of a phospholipid bilayer with embedded proteins (see illustration). Each lipid has both a headpiece and a tailpiece. The head is hydrophilic, meaning that it is attracted to water, and the tail is hydrophobic, meaning that water repels it. If evolution is true and all cells were made by blind, undirected chance, why is the cell structured like that? How would blind, undirected chance know what to let in and what to keep out? That requires information and knowledge.

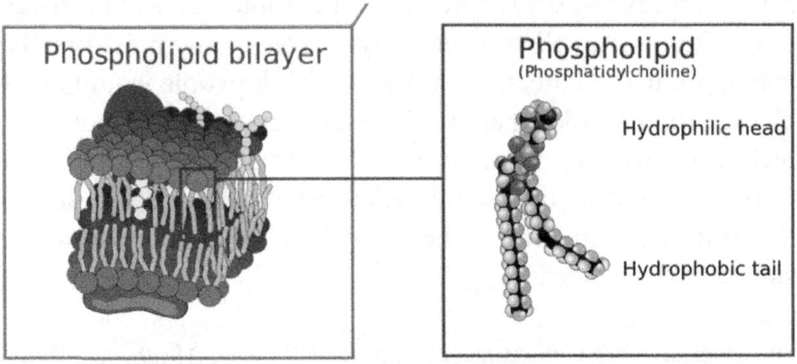

I listened to a medical doctor teaching medical students that the molecules in the primordial soup "arranged themselves by accident until they could transmit information and replicate themselves..." Really? Where did the information originate? How did it get its structure and function? Blind chance can NEVER create exquisite and finely detailed structures like this; that requires intelligence and information. The doctor was just parroting what he was told without questioning the validity of the statement, and so another generation of future doctors is fed the lie. He makes these types of statements throughout the course with a straight face. But he is the authority; how could anyone question him? It is not about what is true; it is about what they tell them to believe is true. Students today are not taught how to think but what to think. With all of our technology and training, we are returning to the Dark Ages, where an elite class of people dictates to the "uneducated" what to believe. Anyone who would dare dissent is treated as a criminal and fired from their jobs. If you doubt that, please watch the Ben Stein movie, *Expelled: No Intelligence Allowed (2008).*

Because of the lipid layer, water can come up to the cell but cannot pass in; it must come through the protein channels. These protein channels have different functions. Some are hormone receptors that signal the cell to perform specific tasks. However, not all cells have receptors for all hormones; it is very specific. Why? How?

You can see in the illustration above that the cell membrane is not like a rubber glove that only protects its contents; it is a very complex structure that is very active. Where did the information come from that tells the cell membrane what to accept and what to reject? Not all cells are the same; each type has its own set of specific receptors, entry, and exit channels. When you realize the enormous amount of activity surrounding the cell, you begin to appreciate how complex and alive it is 24 hours a day.

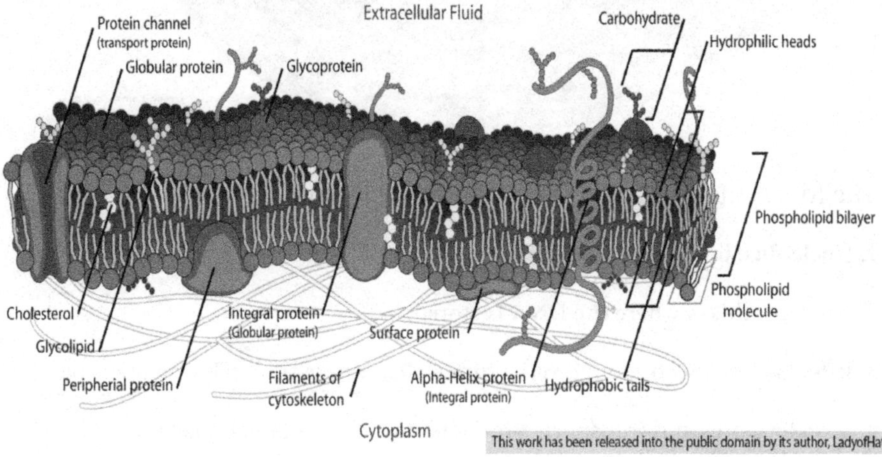

This work has been released into the public domain by its author, LadyofHats

ORGANELLES

Inside a cell are organelles, each having a very specific function to perform. Most organelles have a single phospholipid membrane, but the nucleus has two. Each organelle has a specific job to do in growth and metabolism. There are also specific enzymes associated with each organelle to facilitate the chemical reactions necessary to do its job. The cytosol (the region outside of the organelle but within the cell) has a structure called the cytoskeleton. As the name infers, it is the skeleton of the cell and helps hold its shape. It is also used as a highway for moving things around inside the cell.

The illustration on the next page shows the typical contents of a cell.

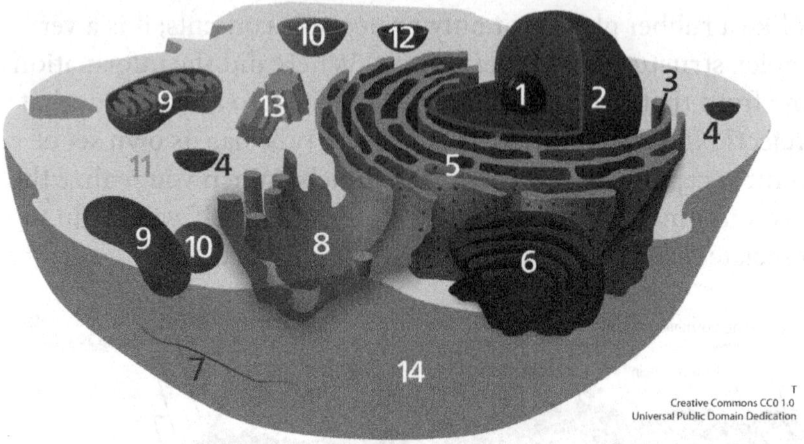

The following is a terse definition of each part:

1. Nucleolus: houses the nucleus

2. Nucleus: this is where the DNA is stored.

3. Ribosomes (dots on rough reticulum walls): where transcription takes place.

4. Vesicle: stores and transports products, and digests metabolic wastes.

5. Rough endoplasmic reticulum: where glycoproteins & lipids are synthesized.

6. Golgi apparatus: packages proteins before they are sent to their destination.

7. Cytoskeleton: the skeleton which helps the cell maintain its shape.

8. Smooth endoplasmic reticulum: (see Rough endoplasmic reticulum)

9. Mitochondrion: where the cell's energy is created.

10. Vacuole: Isolates materials that might be harmful or a threat to the cell

11. Cytosol: the area inside the cell where organelles are found.

12. Lysosome: digests unwanted materials in the cytoplasm (waste disposal)

13. Centriole: micro-tubule-organizing center

14. Cell membrane: the cells outer phospholipid bilayer

What you should realize is that in each of the trillions of cells in your body, there is an active complex factory (something beyond imagination) operating 24 hours a day. Just a few of the activities in a cell are energy production, protein creation, and waste generation and removal. Additionally, 50 to 70 billion cells die each day and are continuously replaced. As a self-contained unit, you can live, breathe, and do things without ever thinking about what is happening at the micro-cellular level. You are a walking miracle! To attribute your life to undirected, serendipitous chance accidents is pure unadulterated fantasy.

Cell types and Tissues

"What you should realize is that in each of the trillions of cells in your body, there is an active complex factory (something beyond imagination) operating 24 hours a day. Just a few of the activities in a cell are energy production, protein creation, and waste generation and removal. Additionally, 50 to 70 billion cells die each day and are continuously replaced. As a self-contained unit, you can live, breathe, and do things without ever thinking about what is happening at the micro-cellular level. You are a walking miracle! To attribute your life to undirected, serendipitous chance accidents is pure unadulterated fantasy.

Cell Types and Tissues

How could undirected, blind chance create and design the different types of cells in the body? We understand now that DNA has all of the instructions for building a body, and epigenetics controls which cells are created in certain parts of the body and, by switching off genes, maintains them. Evolutionary scientists are adept at describing how things happen in the body, and we are learning more each day. However, they cannot determine how or why it happened in the first place. When it comes to explaining how or why, the terminology used changes to words like "may," "might have," "probable" or "probably," "tend" or "tendency," "suppose(d)," and "perhaps." They are very emphatic in saying, "look what evolution gave us," but never explain how it "evolved" in detail. They are emphatic that it happened by accident and that we should just accept it. Listening carefully to their teachings, you can begin to understand why we call it a fairy tale. We will address this subject in more depth in other volumes.

Tissues - amazing design

Epithelial

The epithelial tissues function mainly as protection for the body.

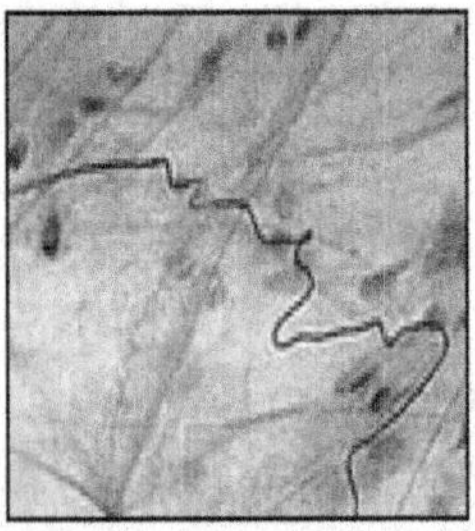

Areolar connective tissue

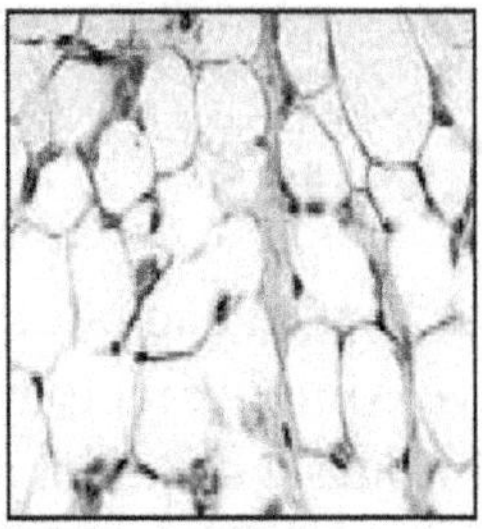

Adipose tissue

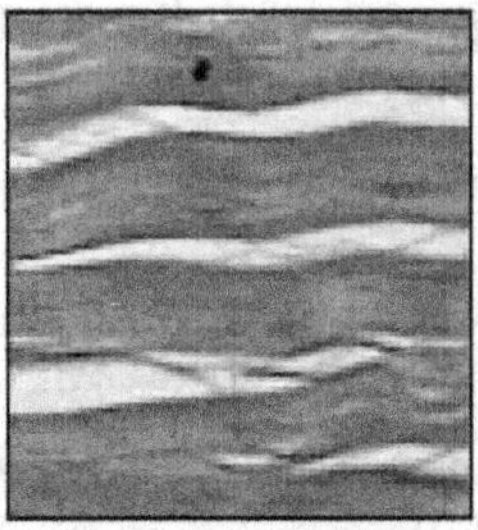

Fibrous connective tissue

Epithelial cells are found in the skin and the lining of the mouth, esophagus, stomach, intestines, bladder, and kidney tubules. In the lungs and blood and lymphatic vessels, these cells help with absorption by diffusion.

Connective tissue

Connective tissue is the most widely used tissue in the body and has a greatly varied use. Some of the uses are connection, protection, insulation, support, and transportation, just to name a few.

One aspect of connective tissue is to hold organs in place and connect the epithelial tissues to other tissues. This tissue is considered "loose."

Muscle tissue

There are three types of muscle cells: skeletal, smooth, and cardiac.

Skeletal muscle - used for locomotion and posture. These cells are under voluntary control by our thoughts. The protein myosin is used for contraction and mobility in muscle and in other parts of the body, along with actin, ATP, calcium, and troponin.

Smooth muscle - These cells are found in the esophagus, stomach, intestines, bronchi, uterus, urethra, bladder, and blood vessels.

Cardiac muscle (myocardium) - these cells are not controlled voluntarily; this is a good design; otherwise, we would have to be thinking about pumping the heart all the time.

The Process of Moving a Muscle

Troponin is used in skeletal muscle and cardiac muscle but not in smooth muscle cells. For an illustrated example, see *Http://www.banyantree.org/jsale/actinmyosin/index.html* by Jeff Sale. Troponin is used to expose the binding site on actin filaments for myosin. The actin and myosin filaments will slide past each other during muscle contraction. The myosin head binds to the actin filament. Magnesium ions cause myosin to release phosphate from the ATP molecule (see the description of ATP in the next section), causing the myosin head to contract. Magnesium and ADP (the result of losing a phosphate molecule) cause myosin to be released from the actin filament, and the myosin stops contracting. Calcium

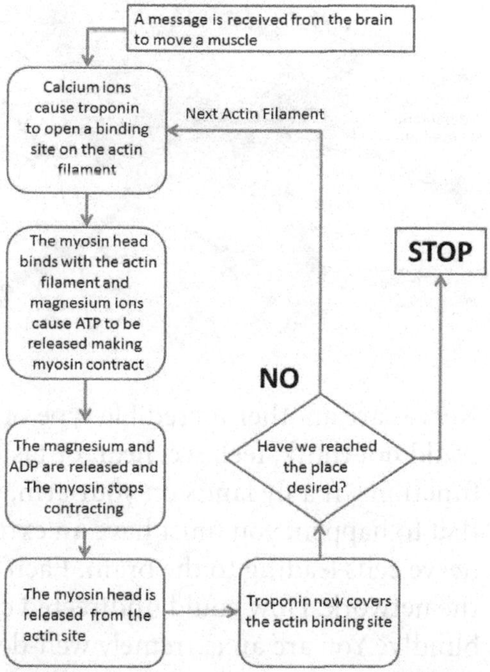

released from troponin covers the actin site, and the process repeats throughout the entire muscle movement. It reminds me of a car jack that moves up by pushing and locking in the next tooth of the jack. It is released from the previous tooth and pushed up to the next. These chemical reactions take place so fast that we are not aware of what it takes to move a muscle. Troponin is used to diagnose myocardial infarctions (heart attacks). Increased levels of the enzyme indicate muscle cell death, and it is released into the bloodstream when there is damage to the heart. This can mean that the heart was simply damaged or that a heart attack is occurring at the moment. After a heart attack, it can take as long as two weeks to return to normal troponin levels.

Nerve cells -

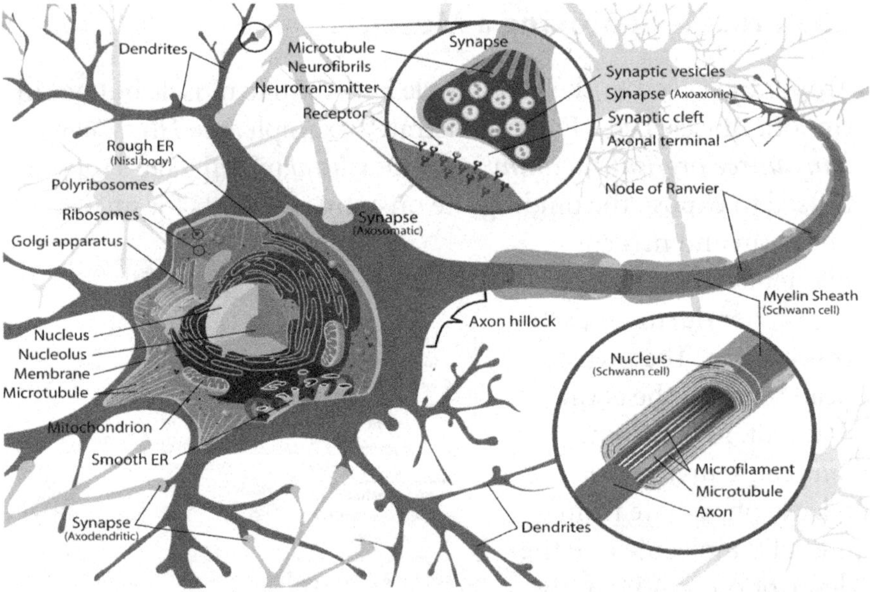

Nerves are another incredible type of cell. Without nerves, you could not think, feel, see, hear, or taste, among some critical functions. If a fly lands on your arm, you feel it immediately. For that to happen, you must have an extremely complex network of nerve cells leading to the brain. Each hair has a connection to the network. How could undirected chance create that network blindly? You are an extremely well-designed machine.

Do Fetuses Feel Pain?

> There is a lot of controversy, currently, over the practice of Planned Parenthood and other abortion providers. The question becomes how soon are nerves developed in a fetus and does the fetus feel pain? Research done by Nicholas Fisk director of the University of Queensland Center for Clinical Research in Australia found that fetuses as young as 18 weeks showed a spike in stress hormones and a shunting of blood to the brain (to protect organs from attack) when an invasive procedure occurred. *[from and article in the New York Times - The First Ache by By Annie Murphy Paul 2/2008]*

Feeling pain is only one reason those who are pro-life are so adamantly o opposed to abortion. The false narrative that fetuses do not feel pain is a blatant misrepresentation of the facts. When the abortionist is crushing the baby or using other hideous procedures after 18 weeks, it is a good bet that the developing baby feels the pain. I can assure you that the Lord of creation hears their cries.

With all the varieties of cells in the body, each having a specific purpose, how would undirected chance ever know that there was a need for a certain type of cell? Can an entity change its DNA? Where did the information come from that built these various cells in the first place? Sure, it comes from DNA, of course, but without a designer, how could these ever have been built in the DNA? Did an accident modify the DNA to define these cells? That is absurd, wishful thinking, and impossible.

ATP synthase - The Power source

Adenosine triphosphate, also is known as ATP is a complex fundamental energy currency for all life. ATP is so critical for life to exist that the lack of it will kill you in a matter of minutes.

Above is a chemical diagram of ATP and ADP. The darker area is ADP the result of one ATP being used.

> "The supply of ATP must be steady because its lack would kill an organism in a matter of minutes. Poisons like cyanide kill so quickly by blocking processing of ATP according to Bergman". *ATP: The Perfect Energy Currency for the Cell by* Jerry Bergman, CRSQ Vol 36(1) June, 1999.

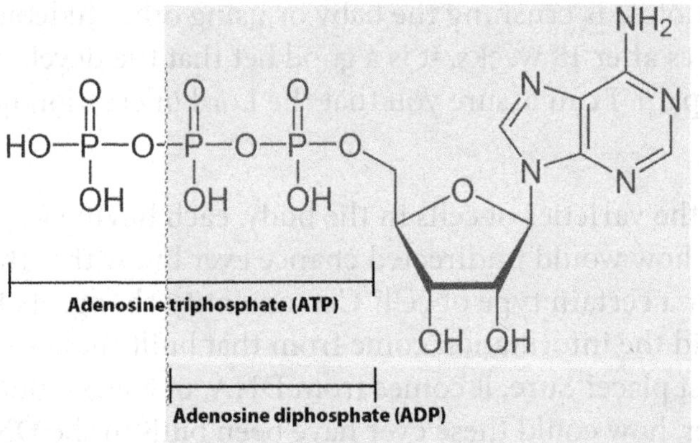

Excuse me if I stop and praise God! Sometimes, I am emotionally overwhelmed by his magnificence. After writing the previous paragraph, I am awestruck. Do you realize how far He is above all others? He is truly worthy of praise. I am perplexed, that anyone would dare suggest that any of this happened by mere chance it is an insult to all thinking humans.

> "One central coenzyme is adenosine triphosphate (ATP), the universal energy currency of cells. There

is only a small amount of ATP in cells, but as it is continuously regenerated, **the human body can use about its weight in ATP per day."** Dimroth P, von Ballmoos C, Meier T (March 2006). "*Catalytic and mechanical cycles in F-ATP synthases. Fourth in the Cycles Review Series*".

The total human body content of ATP is only about 50 grams, which must constantly be recycled every day. The ultimate source of energy for constructing ATP is food; ATP is simply the carrier and regulation-storage unit of energy. The average daily intake of 2,500 food calories translates into a turnover of a whopping 180 kg (400 lbs) of ATP Kornberg, 1989. *For the love of enzymes.* Harvard University Press. Cambridge, MA (Kornberg, 1989, p. 65).

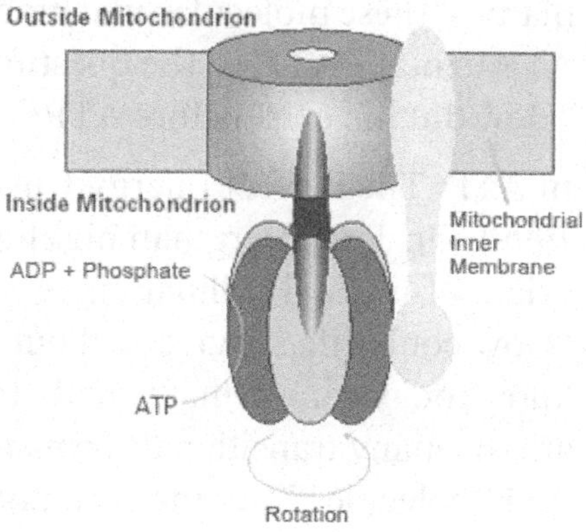

The mitochondria are organelles found in eukaryotic (cells with a nucleus) cells that perform cellular respiration, among other things. Cellular respiration is a series of chemical reactions through which glucose is converted into the energy carrier ATP. The mitochondria have their own genome (mtDNA). This is separate and distinct from the DNA found in the nucleus. This mtDNA originates only from your mother.

There is more going on in the mitochondria, but we will focus on ATP synthase. ATP is generated by converting the ADP molecule (adenosine diphosphate) by adding a third phosphate unit. The ATP synthase has a wheel-like structure that turns up to 200 times per second and produces three ATP molecules on each turn, resulting in six hundred new ATPs per second. Each cell is estimated to have about one billion ATP molecules at any one time. However, this is only enough to last a few minutes, and ADP is recycled to ATP continuously.

You need ATP to move your muscles, among many other things. ATP is not stable enough to be sent to the muscle through the blood, so it must be available at the cell level. It is incredible that there are many of these molecular machines manufacturing ATP 24/7. It takes ATP to make ATP, so the question we need to ask the evolutionist is: "How did life exist before ATP?"

In ATP: The Perfect Energy Currency for the Cell (CRSQ, June 1999), Dr. Jerry Bergman highlights a critical issue regarding the origin of life and cellular energy systems. He poses the questions: "How could life survive without ATP, since no form of life we know today can do that?" and "How could ATP evolve, and where are the many transitional forms required to evolve the complex ATP molecule?" Bergman emphasizes that no viable transitional candidates exist—nor can they exist—because only a fully functional and complete ATP molecule is capable of performing its essential role within the cell. Any incomplete or imperfect form would render the energy transfer system ineffective, making life impossible.

Supporting this view, Dr. David Rosevear, in Interpreting Origins Science, notes that the ATP synthase enzyme—the molecular machine that produces ATP—is itself incredibly complex, composed of thirty-one precisely arranged protein components. If even one part were missing, the enzyme would not function, and the cell would lack the energy required to sustain life. Moreover, both ATP and the ATP synthase enzyme are encoded by DNA, meaning their existence depends on pre-existing genetic information. Together, these observations present a formidable challenge to evolutionary explanations, suggesting instead that ATP and its synthesis system reflect intentional design and irreducible complexity.

ATP is necessary for the creation of ATP and many other processes. Active cell transport (moving molecules in and out of the cell), muscle contraction, the formation of proteins, chemical reactions,

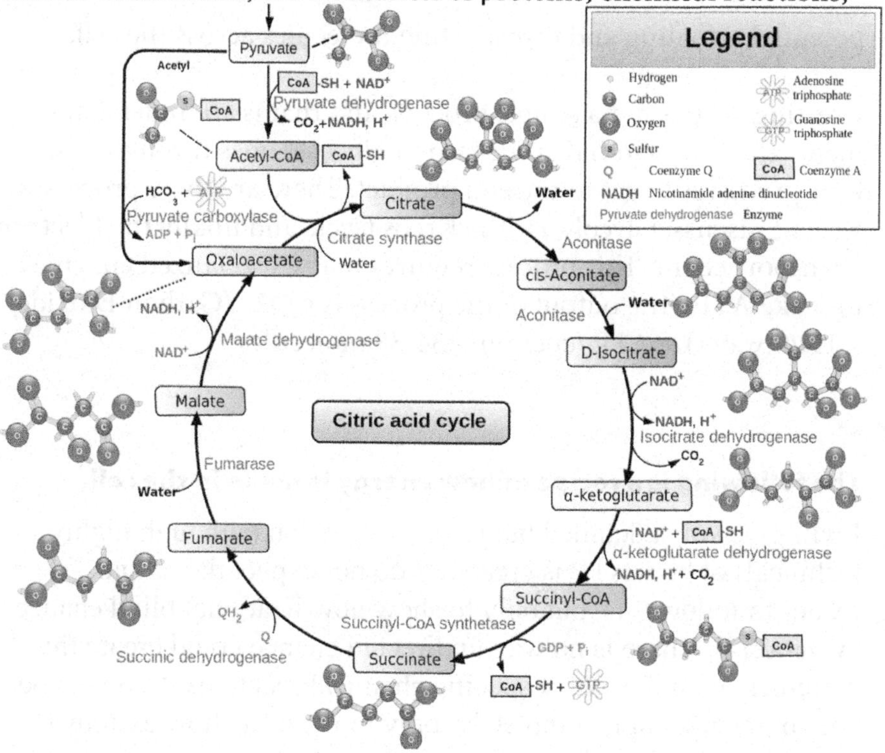

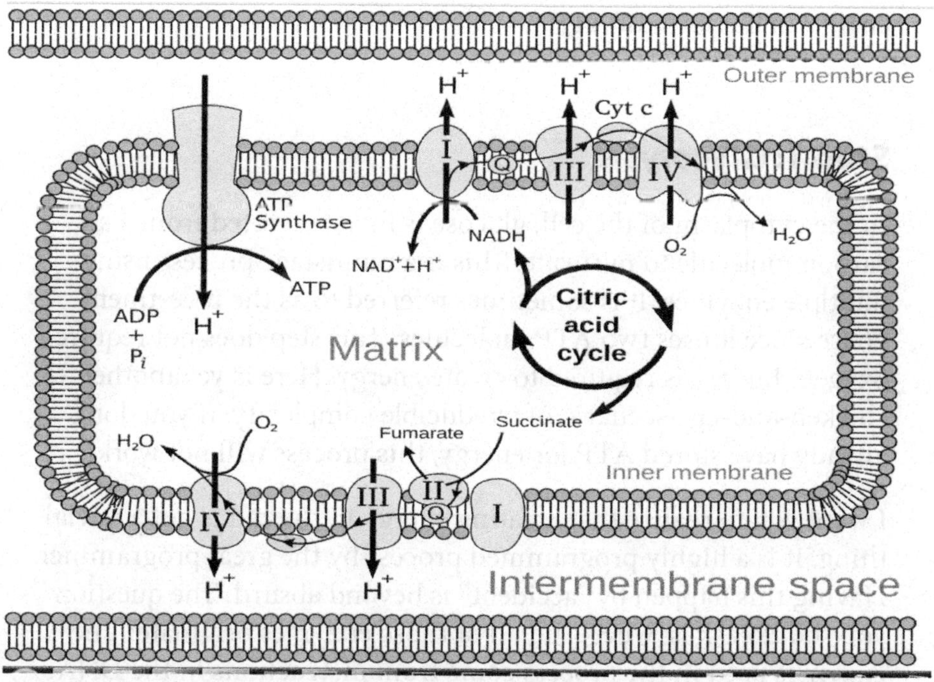

maintenance of the Na/K pump (Sodium-potassium pump), action potential signaling and transporting molecules across the cell.

Creating ATP is not a simple process, and that is an understatement. The raw material (a six carbon glucose) goes through multiple processes to get to the end product. There are three processes we will discuss; Glycolysis, The Krebs Cycle and finally the Electron Transport chain. This process requires many specialized enzymes to make ATP. The output of the process is CO2 (Carbon Dioxide) + H20 (water) + ATP (energy) ~ 34-38 molecules.

The following is a recipe of how energy is made in the cell.

I will give you a detailed but terse explanation (although highly technical) of how ATP is created. I do not expect the casual reader to follow this, but only to show how irrational blind chance evolution is. There is no way undirected chance could create this extremely complex set of specific chemical reactions. I don't expect you to understand it completely; only to illustrate how extremely complex the process is and how many steps there are to creating energy in the form of ATP.

Step 1: Glycolysis

In the cytoplasm of the cell, glucose is first converted from a six-carbon molecule to pyruvate. This is a multistage process using multiple enzymes. It is sometimes referred to as the investment phase since it uses two ATP molecules. This step does not require oxygen. Energy is required to create energy. Here is yet another chicken-and-egg scenario of irreducible complexity. If you don't already have stored ATP for energy, this process will not work.

I want you to realize that creating energy in the cell is not a trivial thing. It is a highly programmed process by the great programmer. Having this happen by "accident" is beyond absurd. The question you should ask is: where did all of those enzymes originate? The enzymes used in the process come from instructions in the DNA. It takes ATP to create ATP, and all of the catalyzing enzymes must be in place for the operation to work. This is an all-or-nothing proposition; either all of the enzymes and ATP are present, or no ATP will be created, and that is the end of life. Small successive steps CANNOT account for this amazing function in the cell. The first phase of this process has ten steps; we must wonder how this process could ever come into being by any other way than design.

The enzyme hexokinase is used to add phosphate to the sugar molecule. The output is glucose-6-phosphate. This step uses one ATP; it then becomes ADP.

The enzyme phosphoglucose isomerase is responsible for the conversion of glucose-6-phosphate into fructose-6-phosphate.

In this step, we use another ATP with the enzyme phosphofructokinase to add another phosphate.

The enzyme fructose-bisphosphate aldolase divides the sugar in half, making two three-carbon molecules. Two molecules are created in this step: GADP and DHAP.

Using triosephosphate isomerase, DHAP is converted to make a second GADP. (From this point on, all reactions are doubled since we are dealing with two molecules.)

Next, hydrogen is removed to form NADH and add a phosphate.

Using phosphoglycerate kinase, two ATPs are created.

Move the phosphate to the middle carbon via the phosphoglycerate mutase.

Two hydrogens and oxygen are removed, and water is formed using the enolase enzyme.

Pyruvate kinase removes a phosphate and forms an ATP. The output of glycolysis is two pyruvates. These pyruvates are moved to the mitochondria.

The output of this step is:

2 pyruvate,

2 NADH,

2 ATP,

2 positive hydrogens, and

2 molecules of water.

The enzyme pyruvate dehydrogenase complex oxidizes it to acetyl-CoA and carbon dioxide. This is a "link" step between glycolysis and the next step. If there is an ample supply of oxygen, then aerobic respiration will take place; otherwise, anaerobic respiration will occur, resulting in lactic acid. Lactic acid is what you get when exercise overwhelms this process (oxygen cannot keep up with demand).

> During intense exercise, the respiratory chain cannot keep up with the amount of hydrogen atoms that join to form NADH, and cannot regenerate NAD+ quickly enough. Wikipedia contributors. "*Lactic acid.*" Wikipedia, The Free Encyclopedia. Encyclopedia, 5 Nov. 2015. Web. 16 Nov. 2015.

Step 2 - The Krebs Cycle

In the link step, Acetyl-CoA (2 carbons) is catalyzed by enzymes from the pyruvate coming out of glycolysis and is now combined with oxaloacetic acid (4 carbons) by citrate synthase, another enzyme, to form citric acid (citrate, 6 carbons).

Using the enzyme isocitrate dehydrogenase, remove two hydrogens to form NADH and oxidize one carbon (CO2).

Add coenzyme A, remove two hydrogens to form another NADH, and oxidize one carbon (CO2).

Remove the coenzyme A and form GTP (later, GTP will be used to make one ATP).

Remove two hydrogens to form FADH2 from FAD embedded in the succinate dehydrogenase enzyme.

Add water, forming an OH group.

Remove two hydrogens to form NADH and create oxaloacetate to repeat the cycle. At the end of the Krebs cycle, from the two pyruvate molecules, we have: 1 NADH x 2 = 2 NADHs (from the link phase) 3 NADH x 2 = 6 NADHs 1 ATP x 2 = 2 ATPs 1 FADH2 x 2 = 2 FADH2s Plus what we gained from glycolysis: 2 ATPs, 2 NADHs.

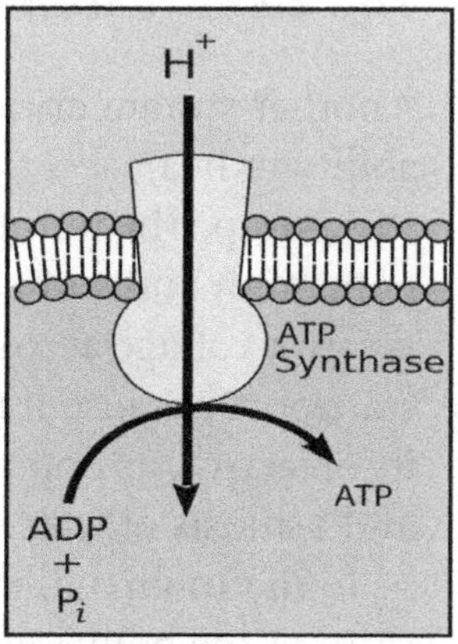

STEP 3 - Electron Transport Chain

The protein complex oxidizes NADH to NAD+ and H+, and some energy is used when electrons are transported to transition molecules. The NADH is at a high energy state. When they are transported, they go from a high energy state to a lower energy state. The electrons are bonded to coenzyme Q and cytochrome C and are eventually reduced to water (H2O).

The energy released from the bonding of CoQ and CytC is used to pump hydrogen protons across the inner membrane of the mitochondria to the outer membrane. You now have a higher concentration of hydrogen molecules in the outer membrane, which creates an electric gradient. Interestingly, the positively charged hydrogen protons want to get back into the matrix; this is when the ATPs are formed. There is a protein called ATP synthase that is used to transport the protons back into the matrix. The ATP synthase has what appears to be an axle, which spins. In the equation, there is an ADP molecule and some phosphate, so as the axle turns, it forces the phosphate to attach to the ADP, making ATP. One NADH will produce three ATPs, and one FADH2 will produce two ATPs. Therefore, the net result of one glucose molecule from all three steps is about 38 ATPs, give or take a few, based on some possible issues during the process. I have given you a high-level description of what happens in creating ATP. Do you still think that blind, undirected "accidents" cause the formation of this amazing process?

The sodium-potassium pump
The Electric Generator

Another system critical to life is the sodium-potassium pump. Your ability to hear, see, think, feel, and move your muscles depends on this pump. The pump is an active transport mechanism for moving potassium and sodium ions in and out of the cell membrane to create a voltage across the cell membrane. It is considered an active transport because it uses energy in the form of ATP to accomplish its objective, as opposed to the diffusion of oxygen, carbon dioxide, and osmosis of water, which are passive transports. The pump can be found in virtually every cell in the body. It is estimated that one-third of the body's energy is used for this process.

Laura Austgen (*Http://www.vivo.colostate.edu/hbooks/molecules/sodium_pump.html*)emphasizes the critical role of sodium pumps in cellular energy consumption by highlighting how extensively they rely on ATP. In resting humans, an estimated 25% of all cytoplasmic ATP is used specifically to power sodium pumps. This demand is even more pronounced in nerve cells, where approximately 70% of ATP is consumed to fuel these pumps. Such figures underscore the essential function of sodium pumps in maintaining cellular homeostasis and the extraordinary energy requirements involved in basic physiological processes like nerve signaling.

In a resting state, the difference in positively charged sodium ions (Na+) on the outside of the cell membrane and positively charged potassium ions (K+) on the inside creates a voltage inside the cell of approximately -70 to -80 mV. It also takes energy to maintain the resting state. A stimulus (electrical or chemical) causes the sodium gates to open in the cell membrane, and the sodium ions begin to move in. This short-circuits the cell, and the voltage drops, creating the action potential (action potentials are also called an impulse or 'spike'). This change in voltage is sent through the nerve and muscle cells in a wave action to neighboring cells and to the destination. These waves only move in one direction because the gates are closed and will only open for a short period. Otherwise, the action would move back in the opposite direction (a wonderful feat of God's engineering). Nerve cells are very poor conductors, so the voltage must be maintained all the way to the destination. Since communication is digital (meaning frequency, not amplitude), the more intense the event (the louder sound, for instance), the more action potentials occur per second. This action potential is an "all or nothing response." If stimulation exceeds the threshold, an impulse will be generated; if it does not meet the threshold, the action potential will not be put into motion.

There are three phases in this process:

1. Depolarization - where the sodium ion gate opens and sodium ions rush in, causing a change in voltage from -70 mv to +20 to +40 mv the inside is now positively charged.
2. Repolarization - the potassium ion gate opens and Potassium rushes to the outside of the cell.
3. Recovery/Refractory Period - sodium ions are now pumped out of the cell and potassium ions are brought back in. Three sodium ions are pumped out for every two potassium ions coming in.

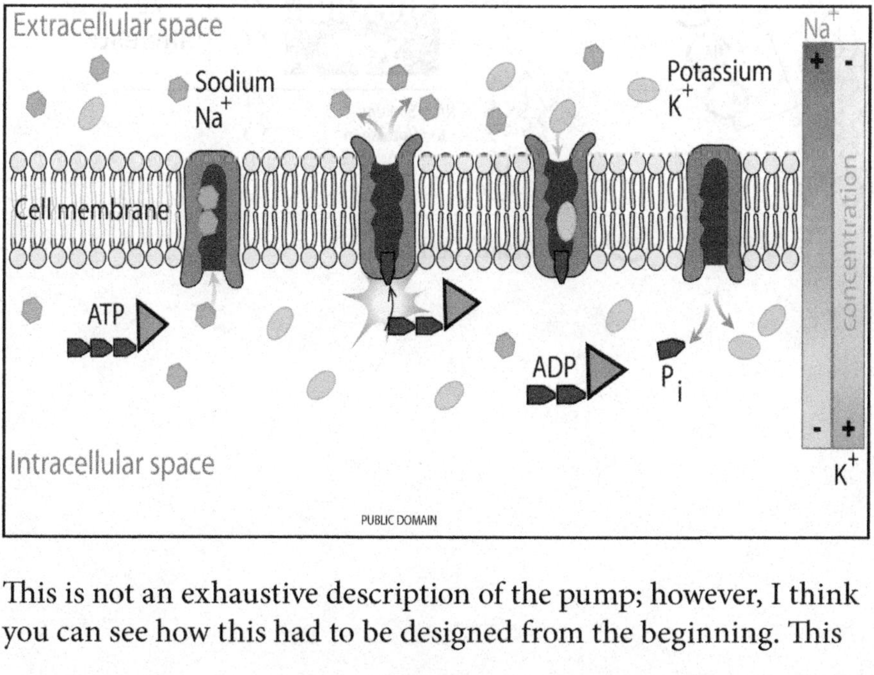

This is not an exhaustive description of the pump; however, I think you can see how this had to be designed from the beginning. This pump must exist, or life in the cell ends, and nerve impulses never reach their destination. The body is a system of collaborating parts that are so interrelated that it would be impossible for these parts to operate on their own. Previously, we mentioned how ATP generates energy for the cell; without ATP, you would quickly die. Sodium-potassium pumps consume a very significant amount of ATP doing their job, showing how critical this function is to life. Without ATP, the body dies; without the electrical energy created by the sodium-potassium pump, the body does not function.

Nerve cells and the Sodium / Potassium pump.

When an event or events happen at or in a nerve cell, this causes an action to take place. However, it must reach a threshold before action is taken. You should be asking how does the nerve cell determine what is enough to cause an action and how would a cell know that?

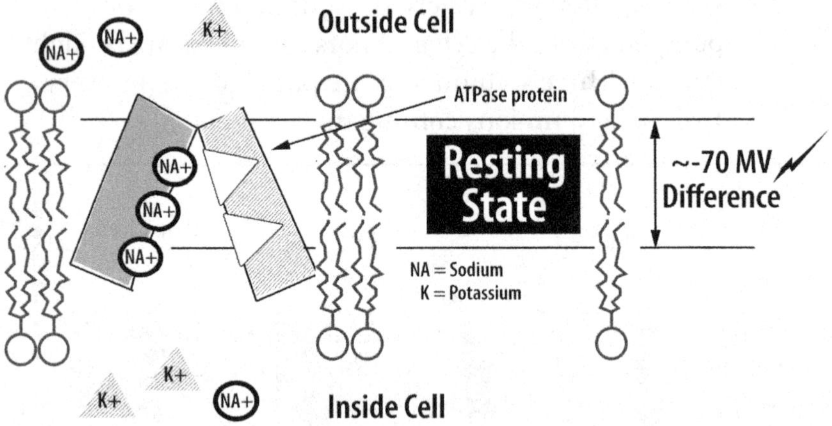

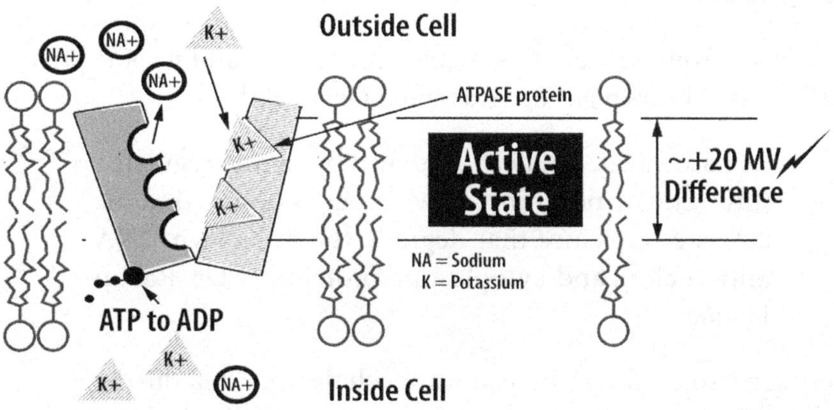

The amazing thing is that God created an insulator over the axon to speed up the signal. This insulator is known as the myelin sheath. The sheath does not have voltage-gated ion channels; this maintains a separation of the charge between the interior and exterior of the cell. However, sodium diffuses along the axon. If the myelin sheath were too long, the signal would be very weak when it reached the next segment. God's design is astonishing; every so often, there is a break in the sheath called the nodes of Ranvier. These are spaced at about every 1 mm or so and contain large amounts of voltage-gated sodium channels, allowing enough sodium into the axon to regenerate the action potential.

In *Anatomy & Physiology: The Unity of Form and Function*, Kenneth S. Saladin explains the role of the nodes of Ranvier in the propagation of nerve impulses. He states that each time an action potential reaches a node of Ranvier, it is restored to its original strength of +35 millivolts. This process ensures that the electrical signal remains strong and consistent as it travels along a myelinated axon, allowing for rapid and efficient communication within the nervous system. The restoration of the action potential at each node is essential for maintaining the integrity of the signal over long distances.

This process of recharging is known as saltatory conduction (from the Latin saltare, meaning to hop or leap). We will go into much greater detail in Book Four concerning the nervous system.

Apoptosis

Apoptosis (from Ancient Greek ἀπό apo, "from" and πτῶσις ptōsis, "fall") is also known as "Programmed Cell Death."

> "A pathway to cell death induced by tightly regulated intracellular programs in which cells destined to die activate enzymes that degrade the cell's own DNA and nuclear and cytoplasmic proteins" - Dr. Rabiul Haque

There are two reasons why you want cell death. First, during embryo development, apoptosis is activated to kill cells between fingers and toes; otherwise, you would have webbed hands and feet. Second, it is used to eliminate damaged cells or cells infected with a virus. This is distinguished from necrosis, which occurs unnaturally due to events like being burned by fire. Cell death in that case is not programmed and not planned. When you are burned, damage occurs that is not repairable, as it is in apoptosis, and a scar will appear. Apoptosis is the process of natural cell death. This is a critical function of homeostasis (maintaining equilibrium) and must work properly, or the results can be catastrophic. There are two ways a cell can self-destruct: internally or externally. If the DNA in a cell becomes damaged, or some other severe cell stress occurs, a set of events takes place that causes the cell to self-destruct (some refer to this as cell suicide). This death is critical to the health of surrounding cells and the destruction of tumor cells. Where does the intelligence come from that tells the cell the DNA is damaged? You have to have design! Undirected blind chance does not have the intelligence to know what to do. There has to be information for this process to exist. Apoptosis is also active in maintaining homeostasis. If cells begin to multiply beyond the required number for optimal health, the apoptosis process is used to reduce the number of cells. If this process is inactivated for some reason, you get uncontrolled cell growth, known as cancer.

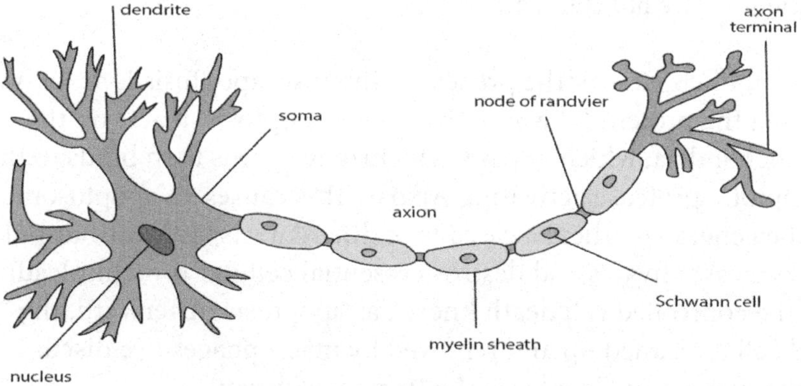

Internal apoptosis: (an overly simplified description)

On chromosome 17, you will find a gene known as the TP53 tumor-suppressor gene; some call it the "guardian of the genome." In normal circumstances, the level of p53 (protein) remains low. However, if an event causes DNA damage or other events that cause cell stress, signals may be sent out to activate the increase of the p53 protein. The tumor-suppressor gene has three functions:

<u>Growth arrest:</u> inhibits the replication of damaged DNA.
P53 binds with another protein, p21; this enters the metabolic process of the cell to shut down the cell cycle.

<u>DNA repair activation.</u>

Apoptosis: if the first two do not resolve the issue.
Chemotherapy and radiotherapy work by activating the p53 gene to fight cancer. However, when examining cancer cells, it has been shown that in over 50% of these cases, a mutated p53 gene is present. This mutation interrupts the cell's natural response to apoptosis, leading to a situation of uncontrolled cell growth (cancer), which can lead to resistance to chemotherapy or radiotherapy. This is why the feedback you want to hear as a cancer

patient is "we got them all."

A simplified view of the process of intrinsic apoptosis is as follows: When the protein BAX is activated for apoptosis, it acts on the mitochondria, which release cytochrome c. This then binds to the apoptotic protease activating Apaf-1. This causes an apoptosome to be generated. Then, a cascade of different caspases (intracellular enzymes) activates and destroys essential cellular proteins, leading to the controlled cell death known as apoptosis. Afterward, the old cell is cleaned up and removed by macrophages (we discuss macrophages in Chapter 6: the Immune System).

External apoptosis: (an overly simplified version)

The second method of apoptosis is through the extrinsic pathway; this method does not use the p53 gene to carry out apoptosis. Through the receptors on the cell membrane known as DR4 and DR5, tumor necrosis factors (TNF-related apoptosis-inducing ligand, or TRAIL) instigate apoptosis. First, TRAIL binds to the receptors DR4 or DR5 on the cell membrane of a tumor cell (normal cells are not affected), and a cascade is set into motion. This sets off the complex known as the death-inducing signaling complex (DISC). Next, inside the cell, the adapter Fas-associated death domain (FADD) is requested; this in turn recruits initiator procaspases 8 and 10. These are then released into the cell cytoplasm, where they activate effector enzymes caspases (executioner) 3, 6, and 7, leading to apoptosis.

Apoptosis is not designed to destroy just old cells, but in embryonic development, it is activated to remove skin between toes and hands. The picture to the right shows a newborn baby's hand, demonstrating complete complex syndactyly of two fingers as a result of incomplete apoptosis. Without this critical gene p53, or the DR4 and DR5 receptors on the cell membrane, cells would generate uncontrollably, and damaged cells would take over in a cascade towards death. One more reason to believe in a designer who knew this could be a factor in life and designed a function to address the issue.

Cell Receptors and Cell Communication

It is fascinating to realize that the body, at the cellular level, is in a constant state of motion. Cells are dying, pumps are pumping, information is being communicated, and energy is being generated while we are awake or sleeping. We are not conscious of all the actions taking place in our body. Let's consider the receptor; a receptor's function is to communicate signals and activate cell functions. There are receptors in the cytoplasm of the cell (inside) and receptors on the cell membrane (outside).

Steroid hormones are small lipid-soluble molecules that pass freely through the plasma membrane. These hormones bind with receptors in either the cytoplasm or the nucleus. We have already mentioned membrane receptors in apoptosis, but there are many other receptors designed to function in specific ways. Because of the function of any particular cell, only those cells that are designed to receive a particular signal will be able to receive it. We can only surmise that during development, the cells were given their function and the receptors they will need to fulfill that function.

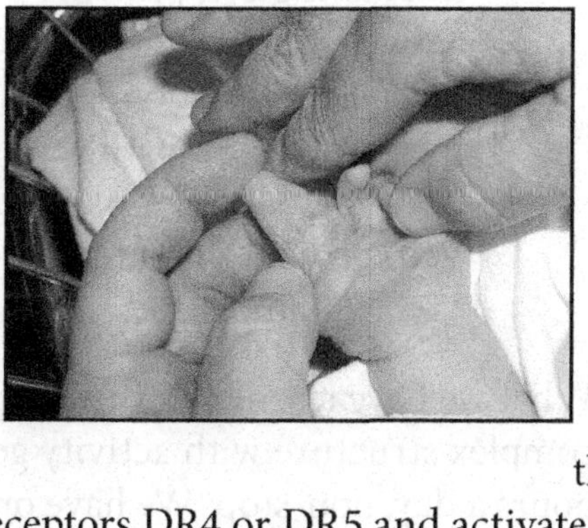

The receptors operate like a lock and key, only allowing the (key) ligand (the ligand is a signal-triggering molecule) to bind to a site on a target receptor (the lock). In apoptosis, we showed that only the TNF (TRAIL) will bind to the receptors DR4 or DR5 and activate programmed cell death.

It is important to understand the process of receptors because it is the basis for the activation of many functions in the living cell. The cell is not just a blob; it is an extremely complex factory that is not only fascinating but awe-inspiring. Once you understand the complexity and the coordination of the huge number of incredible interdependencies in the body, you will appreciate why we say we were designed from the beginning.

Signal transduction pathways diagram (below):

The cell undergoes several reactions to a signal from its receptors. As we have learned, the cell is not a blob that sits idle after its creation. It is a live machine operating in an incredible way 24 hours a day.

The illustration on the previous page is just one example of some of the signaling that takes place. I do not understand how anyone, knowing how complex and designed the body is, can say it is the result of multiple accidents over millions of years. I heard one of the medical professors accidentally say the word "design" and then immediately give credit to evolution. But evolution is blind to design..."

> The activated receptor must first interact with other proteins inside the cell before the ultimate physiological effect of the ligand on the cell's behavior is produced. Often, the behavior of a chain of several interacting cell proteins is altered following receptor activation. The entire set of cell changes induced by receptor activation is called a signal transduction mechanism or pathway. - Wikipedia contributors. "Cell signaling." Wikipedia, The Free Encyclopedia. Wikipedia, The Free Encyclopedia, 19 Jan. 2016. Web. 24 Jan. 2016.

What we have learned in this chapter is that the cell is a very complex structure with activity going on seven days a week, 24 hours a day, non-stop. We have only scratched the surface of what is known about the cell, its inputs, outputs, energy generation, and processing. I have become fascinated by the incredible design of the cell and have been tempted to go into much more detail. If we compare the knowledge available to what we have described here, it would be like comparing the distance from your front door to the moon, and yet we have not opened the door fully. However, it is only necessary to show, from what we can observe, that evolution requires a great deal of faith in a philosophy with very little evidence. The observable evidence shouts vociferously for a designer. Not only are there gaps in the fossil record, but there are also gaps in the evolutionary story, relying mostly on heresy, conjecture, and little white lies.

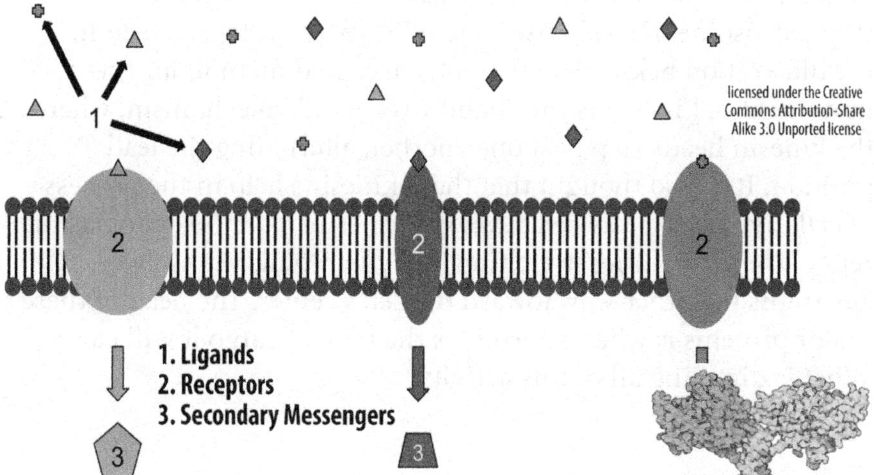

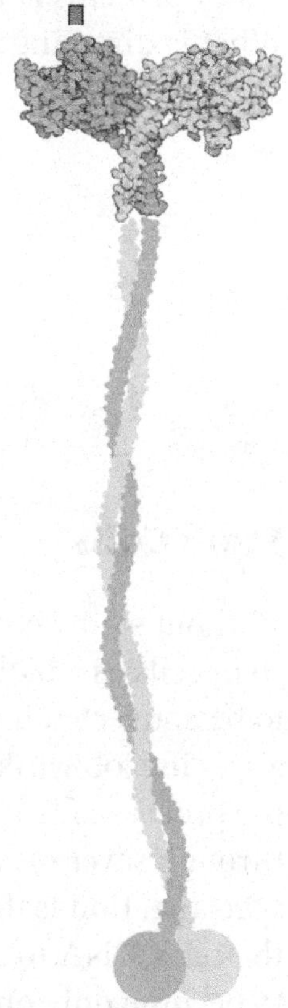

In his review titled "The Inferiority Complex" in the New York Review of Books, evolutionary biologist Richard C. Lewontin makes a striking observation about the behavior of some scientists. He acknowledges that scientists, like anyone else, are capable of telling deliberate lies. What sets these instances apart, he suggests, is the belief that such "small lies" are justified if they serve a perceived "greater truth." This candid admission highlights a philosophical tension in the scientific community—between objective truth-seeking and the temptation to manipulate facts to support broader ideological or theoretical narratives.

Kinesin and Dynein - Transport Movers and Shakers

Another totally amazing function in the cell is the transport of cargo across the cell by kinesin (illustration on the right). Kinesins are proteins that walk across microtubules (the skeleton of a cell), transporting large cargo across the cell toward the plus end of a microtubule (away from the centrosome). Dynein, on the other hand, moves cargo toward the negative end of the cell, toward the centrosome. Kinesin uses one ATP for each step, as seen in the illustration below. The currently accepted method for the movement of kinesin is the "hand-over-hand" mechanism, where the kinesin heads step past one another, alternating the lead position. It is also thought that these kinesins help in the process of cell division. As the cell develops, kinesins pull the endoplasmic reticulum membrane, stretching it. Dynein pulls the Golgi apparatus the other way, toward the cell's center. The head of these motor proteins is what determines the type of cargo it will carry. What is directing all of this activity?

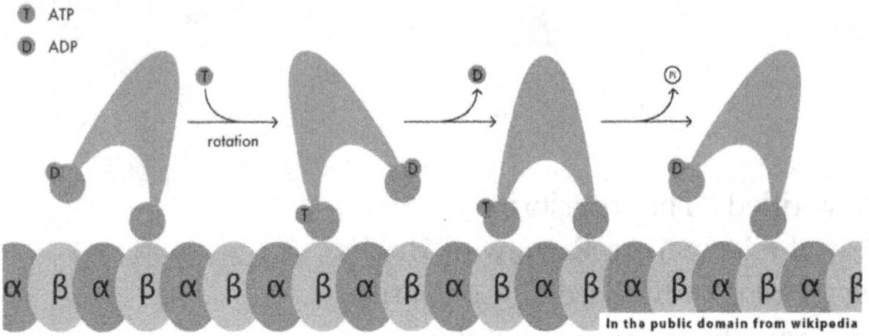

In the public domain from wikipedia

Stem Cells

Without stem cells, you will soon cease to exist. You can think of stem cells as a biological raw material sitting on a shelf waiting to be converted into a specific cell type. Stem cells do not have a specific job while waiting; they are undifferentiated. During the process of becoming differentiated, the stem cell will go through several stages, getting more specific as it develops. This specialization is directed by signals from the body and will cause the cell's DNA to acquire the epigenetic markers that cause it to become only one type of cell. When a stem cell divides, one of the daughter cells is typically left as a stem cell (asymmetric division). This is known as self-renewal, a process that continues throughout your life, replenishing the supply of stem cells. This self-renewal appears to be infinite. However, these cells seem to decrease in number as we age. The other daughter cell can become one of many different types of specific cells. There are three types of stem cells: embryonic stem cells, adult stem cells, and induced pluripotent stem cells. Embryonic stem cells are totipotent (can become any cell type) in the zygote stage. Because embryonic stem cells are totipotent, they are very desirable to researchers trying to find cures for diseases. Then they quickly become pluripotent, capable of becoming any cell in the body but not the placenta and the umbilical cord. That makes sense since the human is being built up from only a few cells. It is a miraculous process whereby a complete human is formed; we will go into more detail in Book II, reproduction. The pluripotent stem cell, like that found in the manufacture of red and white blood cells in bone marrow, can become many types of blood cells: red blood cells and multiple types of white blood cells. In adults, a unipotent stem cell in the skin, for instance, can only become one type of cell. The same is true of other organs known to have stem cells. In 2006, it was discovered that adult stem cells could be reprogrammed to act like embryonic stem cells. This process is known as induced pluripotent stem cells (iPSCs).

In 2006, groundbreaking research by Shinya Yamanaka's laboratory in Kyoto, Japan, revolutionized the field of regenerative medicine with the development of induced pluripotent stem cell (iPSC) technology. Yamanaka and his colleague Kazutoshi Takahashi demonstrated that introducing four specific genes encoding transcription factors into adult cells could reprogram them into pluripotent stem cells—cells capable of developing into nearly any cell type in the body. Their study, titled "*Induction of Pluripotent Stem Cells from Mouse Embryonic and Adult Fibroblast Cultures by Defined Factors*," marked a major milestone in stem cell biology, offering new possibilities for disease modeling, personalized medicine, and tissue regeneration without the ethical concerns associated with embryonic stem cells.

Embryonic stem (totipotent) cells are found in the blastocyst (after a few hundred cells have divided) before it is implanted into the uterus. The issue with these types of stem cells is:

1. They require you to harvest the embryo (not ethical because

it requires a fertilized egg). Life begins at conception, from my perspective. If left alone, that fertilized egg will become a human, barring any issues with the mother or the placenta that cause a miscarriage.

2, These cells are non-autologous (they are not self; see Major Histocompatibility Complex) to the person you are injecting; these will be rejected by the body just like an organ transplant. However, rejection is based on where the cells originate. Brothers and sisters have less of a chance of rejection, while those who are not related have the highest possibility of rejection.

3. It is very difficult to get embryonic stem cells to differentiate into the target cell type you want.

Potency: number of possible fates open to a stem cell.

Totipotent = all fates possible to the stem cell (Zygote)

Pluripotent = Many types of cells

unipotent = 1 or 2 cell types.

Massachusetts Institute of Technology 7.103 Intro to Biology – Hazel Sive

In adults, these cells act as a repair mechanism for replenishing the body. For a developing embryo, the cells differentiate into all the types of cells needed for a growing infant to mature. However, adult tissue-specific stem cells (somatic stem cells) found in the tissue of specific types of cells such as muscle, blood, nerve, liver, teeth, heart, and skin are used to create new cells of that type alone. There are only a few of these stem cells in each area of the body, making it hard to harvest and use; additionally, their number is reduced as we age.

The question is: How do stem cells produce their specific differentiated cell types?

The answer is, today no one knows for sure. In general, these cells divide naturally to keep their numbers up to repair or replace cells as needed. We know, for instance, that injury or disease can cause these cells to start producing their intended cells to quickly repair damaged cells. Stem cells communicate with other cells around them, which is called the stem cell niche. In a process that is still not understood, the cells go where they need to go and become what is called for by the recent communication. There is still much research to be done to understand the process. The hope is that they can find the signal and use it to communicate with stem cells, resulting in repair to damaged tissue. In a broader sense, Transforming Growth Factor Beta, or TGF-β1, a cytokine (used in cell signaling), has many functions in the body, including cell growth, apoptosis, and cell differentiation, which activates stem cells to become a specific cell type.

Regenerative Medicine:

You completely replace the lining of the intestine every four days from stem cells lying beneath the lining! Every bone in the body is replaced every ten years, and the liver is replaced every three years. How does evolution account for this amazing process? Blind chance? Not likely. Unfortunately, at this time, stem cells cannot regenerate a finger or other external body parts. New research on animals that do regenerate limbs has shown that these animals do not have the tumor suppressor genes p53 and p14ARF. By turning off those genes, researchers have found that new cells begin to generate. However, they are a long way off from making this available, but research looks promising. However, turning off the tumor suppressor can have very adverse effects.

In *The Biology of Cancer*, Robert A. Weinberg explains the crucial role of tumor suppressor genes, also known as antioncogenes, in protecting cells from progressing toward cancer. These genes act as safeguards, preventing uncontrolled cell growth by regulating various cellular processes. However, when a tumor suppressor gene undergoes a mutation that results in the loss or reduction of its function, the protective barrier is weakened, allowing the cell to advance along the path to cancer—typically in combination with other genetic alterations. Weinberg emphasizes that the loss of tumor suppressor gene function may be even more significant than the activation of proto-oncogenes or oncogenes in the development of many types of human cancers.

One skin stem cell can reproduce enough cells to cover the entire body. Burn victims now have a way to regenerate skin. By taking some of the stem cells from the burn victim, skin can be grown in a petri dish and then applied to the patient's body. Unfortunately, hair follicles and sweat glands cannot be repaired at this time, so the patient will live but have limitations. A stem cell is a cell of variable potency that can self-renew and is used for organ maintenance and repair.

Hematopoiesis: the manufacturing of blood by the bone marrow.

Red blood cells live about 120 days, white blood cells live only a day or two and must be replaced every day. We need to make 10 billion blood cells every day. The undifferentiated stem cell, called a hematopoietic stem cell (HSC), may go through several steps in its progenitor stage. It will begin its differentiation with the hemocytoblast (hemo = blood, cyto = cell, and blast = they are building something or becoming something specialized) and will become one of many types of blood cells: erythrocyte, basophil, eosinophil, neutrophil, lymphocyte, monocyte, and thrombocytes.

In their 2013 article published in *Pediatric Clinics of North America,* Fernández and de Alarcón describe the developmental progression of the hematopoietic system. Early in development, blood cell formation occurs in the spleen, liver, and lymph nodes. As the bone marrow matures, it eventually takes over as the primary site of blood cell production for the entire body. Despite this shift, certain immune processes—such as the maturation, activation, and limited proliferation of lymphoid cells—continue to take place in secondary lymphoid organs like the spleen, thymus, and lymph nodes. In children, hematopoiesis primarily occurs in the bone marrow of long bones, such as the femur and tibia, whereas in adults, it is largely concentrated in the pelvis, cranium, vertebrae, and sternum. This transition reflects the dynamic nature of the body's blood-forming system throughout life.

The hope for the future of stem cell research is the ability to address diseases for which there is no known cure. If these cells can someday regenerate organs that have failed, we have a bright future. We have mentioned before that transplanted organs fail because they do not contain the major histocompatibility factors (self) in the cells. If you could use your stem cells to regenerate an organ, the issue of "self" disappears.

Adipose-derived cells (ASCs) are cells that are harvested from fatty tissue. It is said that fat has 2,500 times the number of multipotent stem cells (MSCs) as bone marrow. The fat is processed into a mixture of cells known as a stromal vascular fraction (SVF).

In the *November/December 2015 issue of Desert Health*, Drs. Elliot B. Lander and Mark H. Berman discuss the comparative advantages of fat-derived stem cells over those harvested from bone marrow for treating orthopedic conditions. While some orthopedic physicians have promoted the use of bone marrow-derived adult stem cells, research and clinical outcomes suggest that fat, or adipose tissue, provides a superior source. Fat-derived treatments have shown a response rate exceeding 80% for many orthopedic conditions. Additionally, bone marrow extraction is often painful and produces only a small number of stem cells, which must be cultured for several days to reach a usable quantity—typically just a few million cells. In contrast, adipose tissue contains a high concentration of high-quality stem cells in the stromal vascular fraction (SVF), which can be processed without the need for laboratory culturing. With current technology, it is now possible to process fat and extract SVF on the same day, yielding between 10 to 40 million stem cells, making fat-derived sources more practical and effective for regenerative therapies.

According to a February 11, 2015, online issue of *Neurology, the medical journal of the American Academy of Neurology*, stem cells may be more effective than the drug mitoxantrone for people with severe cases of multiple sclerosis (MS). "This process appears to reset the immune system," said study author Giovanni Mancardi, MD, of the University of Genova in Italy. "With these results, we can speculate that stem cell treatment may profoundly affect the course of the disease."

Stem Cells used in curing disease and cancer.

Blood diseases such as leukemia and lymphoma can be treated with just a few multipotent hematopoietic stem cells. This is accomplished by using the patient's healthy stem cells or those from a donor. This sounds simple but, unfortunately, it is not and remains very dangerous, with many possible complications. One potential problem that can be fatal is when the transplant is recognized by the immune system as foreign; it then begins attacking the body, affecting the liver, skin, and other parts of the body. That is not to say that it has not been successful; many have survived after this treatment. If the stem cells come from the patient, there is very little chance of rejection. If the donor is a close relative, there is a slight to moderate chance of rejection. If the donor is not a relative, the chances are good that the process will be rejected. According to The Nation Stem Cell Foundation, statistically, finding a perfect match is less than 25%.

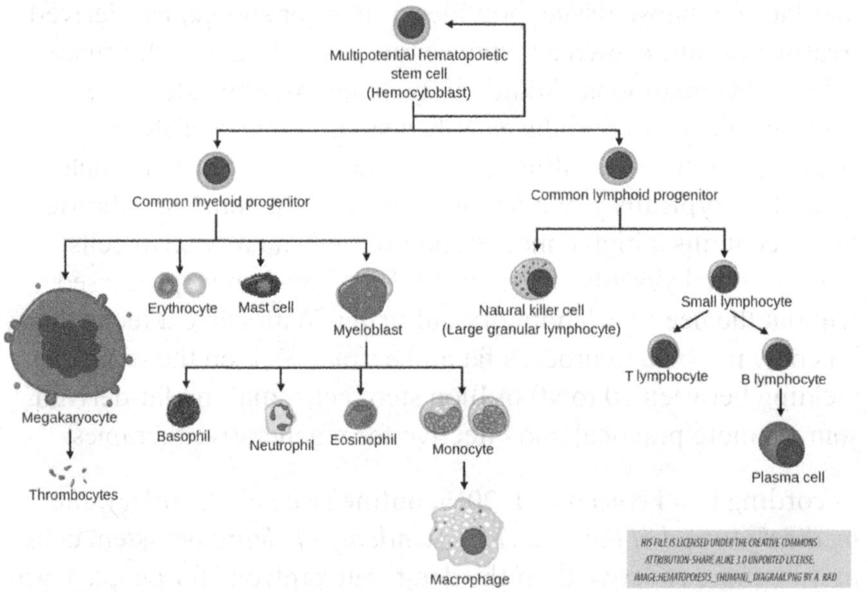

A good site for additional information on actual case studies and success with stem cell therapy is: Http://stemcellrevolution.com. We will address stem cells in more detail in subsequent volumes as they apply to specific parts of the body.

The information given is just the simple basics of the incredible design and activity in the cell. It would take a volume with thousands of pages to fully describe the function of a single cell. In additional volumes of this series, we will address, in more detail, the functions of cells as they relate to specific parts of the body. You may believe whatever you wish about life's origins.
You have two choices, both of which require faith in a higher power:

Faith that natural, undirected chance miracles can explain it all in the face of no observable evidence.
Or
Faith that a great designer planned it all from the beginning, with clear evidence of design everywhere.

I am still puzzled. Evolution requires that you blindly believe that everything was generated by undirected chance, with odds stacked overwhelmingly against it. I have not seen any explanation that describes how non-living material can generate itself into living matter. We learned that non-living material only becomes alive with organization and purpose. How does DNA, an extremely detailed informational structure, self-generate, or how do the immense number of interlinked systems in the cell happen all at once by chance? How do fortuitous mutations cause things to evolve, when observable mutations today are never helpful? Small incremental steps will not suffice; too many systems are interdependent.

Evolutionists are anxious to point out that mutating bacteria become immune to antibiotics. However, we know that some bacteria are already immune for a variety of reasons. One reason is that they do not have a specific protein around them that the antibiotic uses to kill the bacteria. These bacteria are already immune to antibiotics. The "mutation" did not produce a new kind of bacteria or any other creature. Listen to what Pierre-Paul Grassé had to say about the subject.

In *Evolution of Living Organisms*, Pierre-Paul Grassé offers a strong critique of what he calls "present-day ultra-Darwinism," warning that its confident assertions often mislead inadequately informed biologists and foster misinterpretations of biological data. He points to bacteria—organisms that have existed for billions of years—as an example of biological stability rather than evolutionary advancement. Despite minor genetic variations observed in laboratory cultures, bacteria remain confined within their structural framework and show no evidence of true evolutionary change. Grassé draws a crucial distinction between mere variation caused by DNA copy errors and genuine evolution, emphasizing that the two should not be conflated. While bacteria are valuable for genetic and biochemical research, he argues they offer little insight into evolutionary processes. Grassé further criticizes the scientific community for embracing what he sees as a pseudoscience, built on hidden assumptions and bold, often unfounded extrapolations. This, he warns, has taken root in the core of biology and is misleading sincere scientists into believing that foundational evolutionary concepts have been conclusively proven—when in fact, they have not.

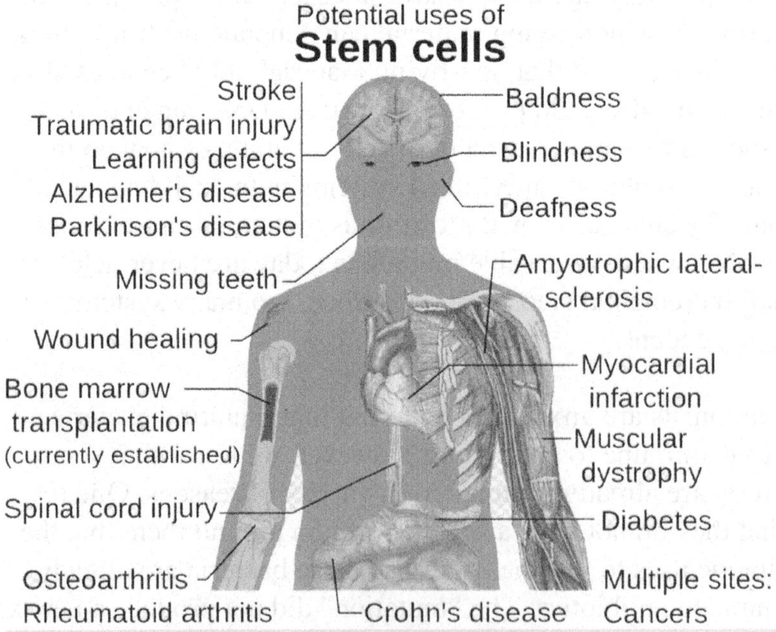

Häggström, Mikael. "Medical gallery of Mikael Häggström 2014". Wikiversity Journal of Medicine 1 (2). DOI:10.15347/wjm/2014.008. ISSN 20018762. - All used images are in public domain.

According to *CreationWiki, The encyclopedia of creation science* The basement membrane is a thin, fibrous layer that lies beneath the epithelium—which covers the surfaces and cavities of organs such as the skin—and the endothelium, which lines the interior surfaces of blood vessels. It is composed of a specific group of molecules found outside the cell, collectively known as the extracellular matrix (ECM). For a long time, these ECM molecules were largely overlooked, but research has revealed their critical role in cellular function. They bind to receptor molecules on the cell surface and transmit signals that influence which genes are activated within the cell. This signaling function highlights the basement membrane's essential role in maintaining tissue structure and regulating cellular behavior.

Dr. Charles Vacanti, professor and chairman at the University of Massachusetts Medical Center and co-editor of Tissue Engineering, emphasizes the significant potential of stem cells derived from liposuctioned fat. These findings are important for several reasons. First, they highlight the effectiveness of adult-derived stem and progenitor cells, which may prove superior to those obtained from fetal sources. Unlike fetal-derived cells, adult stem cells from fat tissue present fewer concerns related to immune rejection and may also be easier to direct into becoming specific tissue types. Most notably, the success of using these adult cells could eliminate the need for fetal cells in therapeutic applications, effectively resolving a major ethical controversy surrounding stem cell research.

You are not aware of the activity that goes on in your body 24 hours a day! It is truly amazing when you begin to realize everything that is going on in the body, and all of that without your direct involvement. You breathe without ever thinking about it. Blood is pumped, oxygen is delivered, and CO_2 is removed. The body's operating system attempts to keep everything in balance, such as blood sugar, body temperature, and many other systems critical to life. Invaders are attacked and eliminated. Cells

STEM CELLS

The Intricate Design of Human Cells Compared to Human-Made Systems

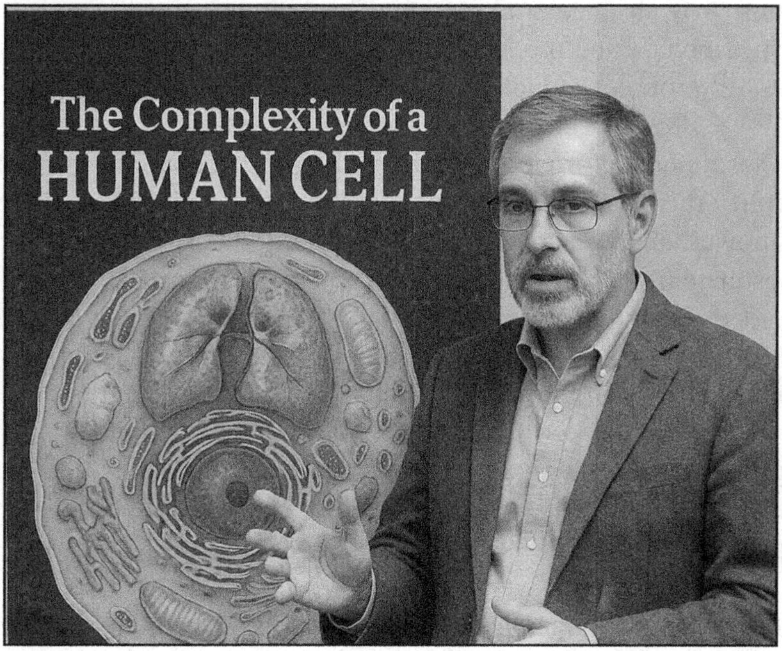

The human cell serves as the fundamental unit of life, displaying remarkable complexity and organization that is essential for the maintenance of homeostasis and the overall functioning of the organism. Each cell operates through a network of coordinated systems, where a complex interaction between various organelles and molecular components ensures the survival and efficiency of cellular processes. To understand the depth of this complexity, it is necessary to consider the interactions in layers and feedback mechanisms that characterize cellular behavior. As thoroughly explained by Ma'ayan (2017), systems biology offers information about these dynamic relationships, revealing how signals, pathways, and various cellular activities intertwine to form a cohesive entity.

In the center of cellular complexity lies the notion of irreducible complexity, a principle that postulates that certain biological systems are composed of interdependent parts that work only when they exist together. For example, the cellular machinery involved in protein synthesis—comprising ribosomes, transfer RNA (tRNA), messenger RNA (mRNA), and several enzyme factors—illustrates this phenomenon. Each component plays a crucial role in a series of complex steps that transmit genetic information to functional proteins. Any absence or alteration of a part interrupts the entire system, emphasizing the need for a complete assembly for successful protein production.

Energy-supporting processes that depend on mitochondria further exemplify this principle. Mitochondria are often referred to as the cell's powerhouses, responsible for the production of adenosine triphosphate (ATP) through oxidative phosphorylation. However, this process requires a variety of finely tuned components, such as the electron transport chain and ATP synthase. The interruption of any single element within this network results in a decreased ATP output, thus affecting the cellular functions that depend on energy. Therefore, the total dependence on organized sets that must collaborate synergistically affirms the notion of irreducible complexity in human cells.

By comparing these biological structures with man-made projects, it is evident that similar principles govern engineered systems, especially within the fields of robotics and complex machinery. For example, consider a car mechanism, which integrates multiple components, including pistons, valves, and fuel injectors, all working collectively to produce movement. The absence or malfunction of a single part can lead to engine failure, mirroring the consequences observed in cellular systems in which functionality depends on the integrity of their components.

Intentionality in biological structures is also parallel to the orchestrated design found in human engineering. Each organelle within the cell operates as part of a larger system, displaying efficiency and purpose that reflects an underlying plan similar to architectural design. The presence of specific structures such as the endoplasmic reticulum, the Golgi apparatus, and lysosomes, serves different functions analogous to organized compartments within a manufacturing process. The design of these cellular systems implies a forethought level, suggesting that the design that shaped these structures have led to non-random arrangements, but rather to optimized configurations that increase viability.

This exploration of the complex architecture of the human cell not only emphasizes the intricacy of its operations but also highlights the implications of intentionality that can be extracted from its design. Examining interconnectivity and irreducibility at the cellular level provides deeper insights into the need for a coordinated system that parallels man-made buildings and emphasizes the sophistication inherent in biological life. The human cell is a notable unit of biological organization, which contains several specialized structures known as organelles, each contributing to its maintenance functions of life. The nucleus, mitochondria, and ribosomes are among the most critical components that support cellular functioning. The nucleus serves as a control center, housing the genetic material of the cell (DNA) and regulating gene expression, which in turn influences cell activity and differentiation (Alberts et al., 2015). The mitochondria, often called the powerhouses of the cell, are responsible for the production of adenosine triphosphate (ATP) through oxidative phosphorylation, allowing various cellular processes to occur (LANE, 2015). The ribosomes function as the site of protein synthesis, translating messenger RNA (mRNA) into polypeptide chains that fold into essential functional proteins for cellular operation (Stark et al., 2018).

Irreducible complexity, a term popularized by Michael Behe, describes systems composed of multiple interdependent parts, where the elimination of any unique component results in the system ceasing to function (Behe, 1996). This concept is vividly illustrated in the functioning of the human cell. For example, intricate machinery involved in the generation of ATP within mitochondria cannot function without the sequential actions of various proteins and enzymes, non-redundant mitochondrial DNA (mtDNA) that encodes essential proteins, and a functional membrane. The interruption of any part of this complex system leads to a failure in energy production, affirming that the coordinated interaction of all elements is indispensable. In the same way, the ribosomal function is based on the correct ribosomal RNA (rRNA) and numerous ribosomal proteins, which underlines the need for each component to achieve the overall objective of protein synthesis.

The interdependent nature of cell organelles exemplifies coordination reminiscent of human engineering systems. Consider the modern car, which integrates multiple systems (motor, transmission, electrical systems) that must work in harmony to ensure that the vehicle operates efficiently. The failure of any unique component can result in deteriorated performance or total failure, reflecting irreducibly complex systems observed in biological cells (Dembski, 2010). In addition, it parallels how engineers design components with specific roles and interactions to meet a functional objective. Evolution through natural selection raises interpretive challenges to elucidate such master complexities present within cell structures, often leading to discussions about intentionality in biological designs (Meyers, 2018).

For example, the interaction between the endoplasmic reticulum (ER) and the Golgi apparatus highlights a concerted effort to handle protein folding, modification, and transport. The ER serves as a site for initial protein synthesis, and incorrect folding can lead to cellular stress or disease, which is rectified by cellular quality control mechanisms (Miller et al., 2017). The Golgi apparatus, in turn, modifies and classifies the proteins received from the ER, preparing them for their designated destinations, reinforcing the notion that the functionality of these organelles is interconnected in a highly organized manner.

This deep synchrony within cell structures raises profound questions about the origins of such complexity. The deliberate functional design parallels those found in sophisticated human manufacturing systems, indicative of an intentional disposition rather than a mere product of random mutations and natural selection. The deliberate assembly of proteins, organelles, and pathways within the cell resembles deliberate engineering processes, suggesting that such complexities may not simply indicate an evolutionary advantage but can also provide insights into the underlying intentional structures within biological phenomena. The comparative study of cellular systems and advanced engineering systems underscores the integrated sophistication within human cells, inviting greater exploration into the implications of such complexity in our understanding of origins and the organization of life. The irreducible complexity states emphatically that some biological systems cannot function if a part is removed, underlining the intentional design underlying these structures (Behe, 2021). This notion extends to various systems within human cells, in which the interactions and interdependencies of the molecular components reveal sophisticated mechanisms that are fundamental to cell function. A salient example can be found in the signaling pathways of the cell that mediate a multitude of cellular responses to environmental stimuli. These pathways often involve a series of orchestrated processes with precision, which include alterations in the conformation of proteins and sequential phosphorylation of target proteins. Each element within these pathways is essential; the absence of a singular component can lead to a complete breakdown of the signaling process (Davis & Sumara, 2014).

For example, consider the mitogen-activated protein kinase (MAPK) pathway activated by mitogens, which transmits extracellular signals through a cascade of phosphorylation events that lead to cell proliferation, differentiation, or apoptosis. The integrity of this pathway is based on a network of protein interactions, where the loss of any part—from the receptor that initially captures the extracellular signal to the transcription factors that ultimately guide gene expression—prevents the pathway from transmitting its message. Therefore, the MAPK pathway exemplifies irreducible complexity, highlighting the interdependence of its individual components and their coordinated functionality (Davis & Sumara, 2014).

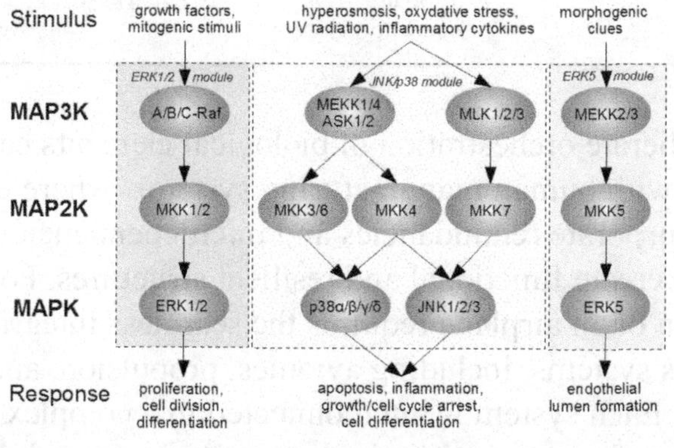

The intricate nature of cellular structures contributes to this sense of irreducible complexity. For example, consider the bacterial flagellum, often cited as a distinctive example of irreducible complexity. The flagellum comprises numerous proteins that assemble into a rotary motor, enabling bacterial motility. Each protein in this system plays a critical role; the absence of a single filament or motor protein entirely compromises the flagellar function. Behe (2021) clarifies how this complexity challenges the gradual picture of evolution, stating that the totality of the flagellum's structure indicates an underlying purpose.

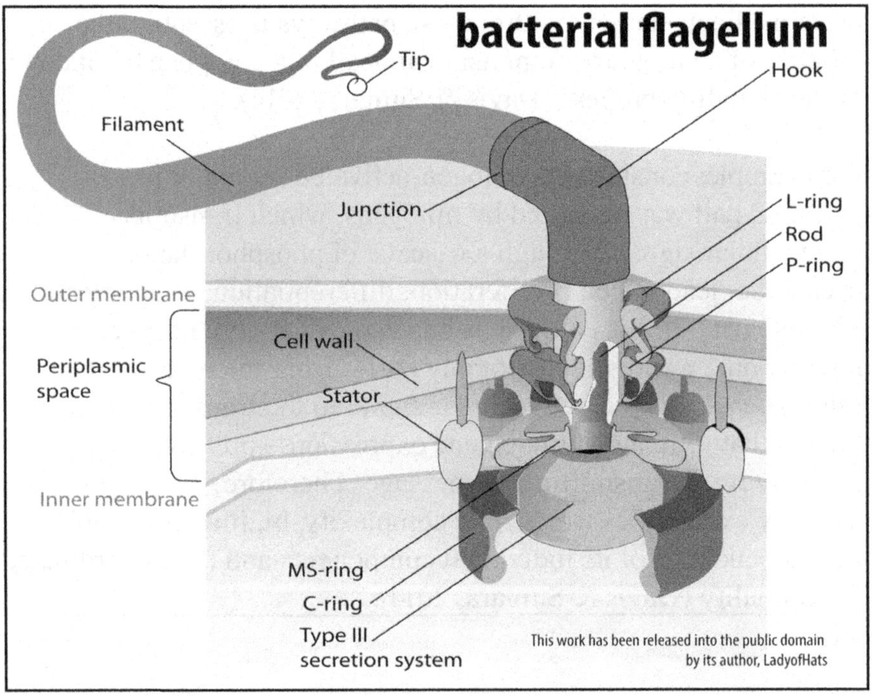

This deliberate orchestration of biological elements can also draw parallels with human manufacturing systems, where engineers often incorporate redundancies and interdependencies into the design to create functional and resilient structures. For example, the design of an airplane requires the seamless integration of numerous systems, including avionics, propulsion, and structural integrity. Each system is interconnected in a complex manner; the failure of a part, such as in the event of engine failure, can compromise the entire safety and functioning of the aircraft.

The similarities between these engineered systems and cellular structures underline a shared model of intentionality, clarifying the principles that govern complexity in both biological and human-designed systems.

Furthermore, the mapping of complexity extends to broader scales, where Thomas and Zaytseva (2016) argue that human knowledge itself can be viewed as a complex adaptive system. From this perspective, knowledge, much like cellular function, is not simply a sum of its parts but an interconnected network of information that depends on the relationships between different concepts and ideas. Each piece of knowledge contributes to a broader understanding, similar to how the functionality of a cell emerges from the interaction between its molecular components. This comparative analysis highlights the intentionality in design, both in biological systems and in human engineering, suggesting that complexity may not arise by chance but rather ***through intentional structure and function***.

The intricate operations of human cells can be compared to sophisticated human manufacturing systems, revealing a compelling semblance of order and design. One of the most important characteristics of cellular function is communication, which operates through a series of highly coordinated molecular signals. This system of biocommunication exhibits surprising similarities with communication protocols used in modern engineering systems, such as telecommunications and network infrastructures.

Cell communication relies on the exchange of signals through various pathways, reminiscent of data transmission in engineered networks. For example, signaling pathways, such as the MAPK/ERK pathway, operate in a highly specified manner in which each phase (activation of the kinase, substrate binding, etc.) must occur sequentially for the appropriate cellular response to be executed. This biological complexity is analogous to layered protocols in computer networks, wherein data packets must follow a defined path to ensure adequate delivery and processing.

Klüver et al. (2022) also illustrate that the principles underlying biological systems can be reflected in human manufacturing projects, highlighting how cellular networks achieve function through redundancy and robustness in their responses.
Cells distribute a variety of signaling molecules, including hormones, neurotransmitters, and cytokines, akin to multi-node communication in engineered systems, where different pathways ensure that signals are transmitted efficiently, even in the presence of potential interruptions. For instance, the ability of cells to adapt their responses to fluctuations in environmental conditions can be compared to adaptive algorithms in technology that adjust to maintain optimal performance.

Additionally, cellular structures exhibit what is termed "irreducible complexity," the idea that some biological systems cannot be simplified further without losing their functionality. Taking the bacterial flagellum as a case study, this structure necessitates multiple interacting proteins to function as a cohesive unit for motility. This contrasts with human engineering, where complex machinery, such as a jet engine, relies on numerous interdependent components to achieve flight. Just as a plane cannot function without its engine, control systems, and structural integrity, a cell with a malfunctioning flagellum cannot achieve movement.

The complexities observed in the provision and orchestration of cellular components invite consideration of the intentionality behind their design. Biomolecular machinery, such as ribosomes, performs the critical function of protein synthesis, demonstrating a deliberate orchestration akin to assembly lines in manufacturing plants. These ribosomes coordinate the sequential addition of amino acids to form polypeptides, illustrating a precision that reflects planned engineering processes. The regulatory phases involved in gene expression further exemplify this intentional design, wherein intricate mechanisms operate with a specificity that parallels mechanical systems guided by precise engineering principles.

Moreover, the cellular economy of resources emphasizes the intentionality in design as seen in human systems. For instance, the maintenance of homeostasis within human cells showcases a complex interaction between various systems, resembling efficient designs from an energy perspective in engineering. The mutual feedback in cellular metabolism, through pathways such as glycolysis and the citric acid cycle, ensures that energy is effectively allocated based on cellular needs, similar to the optimization of resource distribution in integrated circuits.

Ultimately, the parallels between cellular systems and human-engineered projects underline not only the complexity of life at the cellular level but also the intentionality that governs biological structures, inviting deeper investigation into the nature of existence and innovation. The notion of intentionality in biological systems arises as a fundamental theme in contemplating the complexities inherent in human cells. Unlike simple structures, the human cell manifests an elaborate interaction of functions and components that exhibit characteristics of design. As articulated by Forestiero (2022), the historical development of these systems suggests a trajectory shaped by purposeful interactions rather than a mere sequence of random events. Each element within the cell, from organelles to the cell membrane, operates within a highly regulated context, where different components are essential to the overall functionality of the cell.

By framing biological complexity through a lens of intentionality, it is essential to recognize how complex and coordinated the systems within the cell are. For instance, consider protein synthesis processes and cellular respiration, which involve multiple interdependent molecular machines, including ribosomes, mitochondria, and various enzymes. Each component has a defined function that cannot be transferred or omitted without compromising system integrity. This interconnectivity exemplifies the concept of irreducible complexity, where the parts are interdependent in such a way that the absence of any component fundamentally interrupts the operation of the system (Behe, 1996).

Vaage (2020) further extends this idea by comparing biological systems to engineered machines. Both domains exhibit a remarkable ability to leverage resources and perform complex tasks through organized structures. For example, the sophisticated architecture of the cytoskeleton can be likened to the structural design of a bridge, where each filament provides support and facilitates material transport. This analogy emphasizes that biological and artificial systems must be deliberately designed to achieve functionality. The specificity found in cellular mechanisms compares them to the intentional projects observed in human inventions, suggesting that the origin of such complexity may not be solely the result of chance events but rather an intentional structure.

Cellular regulation processes, including gene expression and cell signaling networks, serve as additional evidence of intentionality. The orchestration of gene regulation involves not only direct interactions of various transcription factors but also layered responses to environmental stimuli. A highly regulated system is similar to programming protocols in computer systems, where specific inputs yield definitive outputs. Thus, both biological network systems and human technological environments incorporate intentional choices made within their respective contexts, highlighting ***the idea that complexity arises from design principles rather than random occurrences.***

Moreover, the intriguing comparison of cellular systems with human engineering projects invites a deeper exploration of purpose in nature. Just as engineers implement predefined constraints and objectives in the design of machines, the processes observed in human cells reflect intrinsic goal-directed behavior. This can be illustrated through the efficiency of cellular metabolism, where energy conversion processes are optimized for survival, functionality, and adaptation. By examining biological mechanisms through this lens of intentionality, it becomes evident that complexity within cells is likely the result of deliberate configurations, similar to the systematic approaches observed in human-made technologies.

Thus, the interplay between biological intricacies and engineering projects highlights a shared narrative of intentionality that transcends the dichotomy of nature and nurture. Far from being a consequence of chaotic processes, the functioning of the human cell illuminates a structured sophistication that resonates with the principles of design evident in the influence of human beings on creations. Through this lens, the phenomenon of biological complexity can be appreciated as a manifestation of purpose that permeates the cellular landscape. The intricate workings of the human cell, characterized by a myriad of interdependent systems and components, provide a profound illustration of irreducible complexity. This notion postulates that certain biological structures cannot be simplified without losing their essential function. The human cell operates through a wide range of organelles, each performing specific but interrelated tasks, such as energy production through mitochondria, protein synthesis through ribosomes, and genetic information processing within the nucleus. The absence or dysfunction of any unique component can lead to systemic failure, underlining the principle of irreducible complexity as it applies to cell biology.

When side by side with systems designed by humans, the similarities in complexity and coordinated functions are striking. For example, engineering systems, such as computer networks or automated manufacturing facilities, exhibit complex interactions among their components, similar to multilayer interactions inside a cell. In both contexts, the intentional disposition and interdependence of the parts are critical for the system to function effectively. This comparison raises significant questions about the nature of biological systems. As Dembski (2013) highlights, the presence of intricate designs in both organic and synthetic domains often implies the existence of a guiding intellect, a notion suggesting an underlying intentionality in the structural arrangements observed in biology.

The implications of such biological complexity extend far beyond mere academic discourse. As Van Regenmortel (2019) points out, leveraging this understanding can lead to advancements in several fields, particularly medicine and biotechnology.

This exploration into the complexities of cellular mechanisms not only enriches our understanding of life itself but also highlights the shared design principles found in both nature and human creations. The parallels drawn between cellular systems and human engineering can serve as the foundation for interdisciplinary research, promoting collaboration between biologists, engineers, and information theorists. The intentionality observed in biological structures prompts further investigation into the principles governing life and the extent to which complexity can be replicated or optimized through human ingenuity.

As we delve deeper into the layers of cellular functions, we are likely to uncover new knowledge boundaries regarding origins. Future studies can capitalize on advancements in microscopy and computational modeling to reveal the dynamic interactions that characterize cellular behavior, paving the way for transformative advancements in our understanding of the intricate orchestration of life. Continuous interaction between the pursuit of knowledge and the application of findings in practical contexts promises to reformulate our approach to health and technological challenges.

In conclusion, the intricate workings of the human cell provide a compelling case for examining the broader implications of complexity in biological systems, ultimately encouraging a multidisciplinary approach that encompasses scientific research and innovative design thinking.

CITATIONS

Van Regenmortel, Marc HV, and Marc HV Van Regenmortel. The rational design of biological complexity: a deceptive metaphor. HIV/AIDS: Immunochemistry, Reductionism and Vaccine Design: A Review of 20 Years of Research (2019): 87-102. Https://link.springer.com/chapter/10.1007/978-3-030-32459-9_8

Ma'ayan, Avi. Complex systems biology. Journal of the Royal Society Interface 14, no. 134 (2017): 20170391. Https://royalsocietypublishing.org/doi/abs/10.1098/rsif.2017.0391

Behe, Michael J. The Argument from Biological Complexity. Contemporary Arguments in Natural Theology: God and Rational Belief (2021): 89-108. Https://www.torrossa.com/gs/resourceProxy?an=5203011&publisher=FZ0661#page=100

Thomas, John, and Anna Zaytseva. Mapping complexity/Human knowledge as a complex adaptive system. Complexity 21, no. S2 (2016): 207-234. Https://onlinelibrary.wiley.com/doi/abs/10.1002/cplx.21799

Klüver, Christina, Jürgen Klüver, and Jörn Schmidt. Modeling Complex Processes Through Nature-Analogous Methods. Https://link.springer.com/content/pdf/10.1007/978-3-658-46510-0.pdf

Ewert, Winston. Digital irreducible complexity: A survey of irreducible complexity in computer simulations. BIO-Complexity 2014 (2014). Https://bio-complexity.org/ojs/index.php/main/article/viewArticle/81

Forestiero, Saverio. The historical nature of biological complexity and the ineffectiveness of the mathematical approach to it. Theory in Biosciences 141, no. 2 (2022): 213-231. Https://link.springer.com/article/10.1007/s12064-022-00369-7

Vaage, Nora S. Living machines: Metaphors we live by. NanoEthics 14, no. 1 (2020): 57-70. Https://link.springer.com/article/10.1007/s11569-019-00355-2

Davis, Brent, and Dennis Sumara. Complexity and education: Inquiries into learning, teaching, and research. Routledge, 2014.

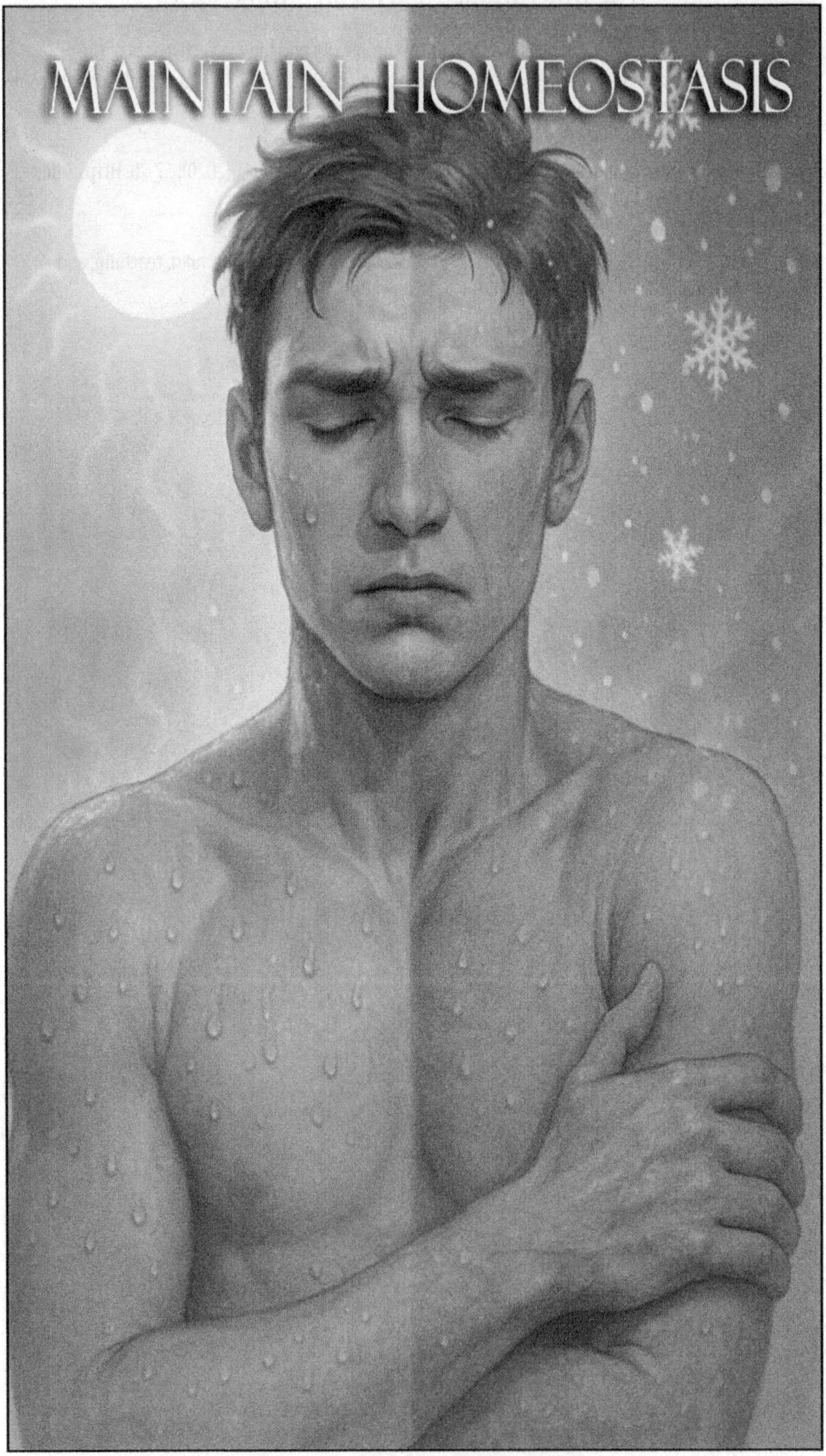
MAINTAIN HOMEOSTASIS

5

Homeostasis

Keeping the body operating as it was designed.

Homeostasis

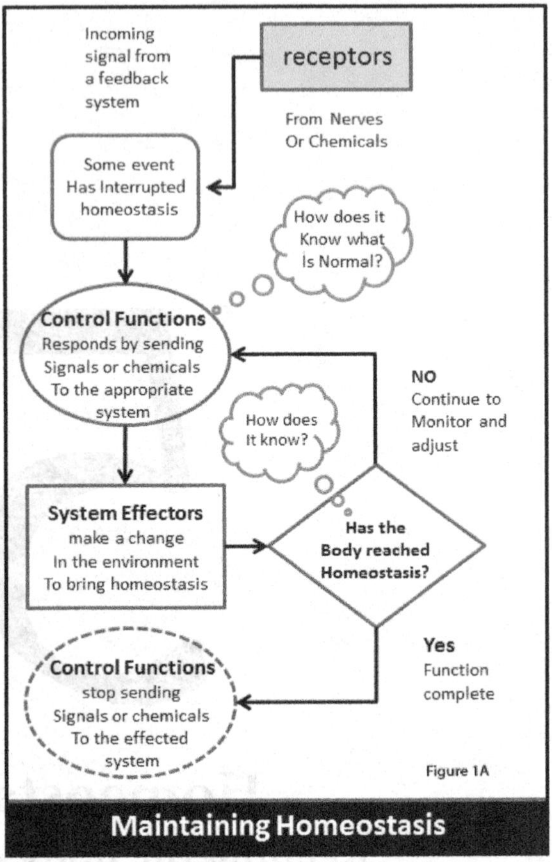

Your body works best when systems are kept within a very small range of tolerances or set points. There are sensors that relay messages to different glands, which in turn send messages to various parts of the body to take action to bring the body back into balance. Homeostasis, derived from the Greek meaning "same" and "steady," is essential because things change in the environment in which we live. The body must have a way to keep systems in balance, such as body temperature. One amazing aspect of the human body is its ability to maintain homeostasis. Homeostasis maintains optimal health through multiple feedback systems. These systems are constantly monitoring, evaluating, and reevaluating to maintain balance. Without this function, the body would certainly die.

Evolution makes no sense for the simple reason that undirected chance and natural selection can never explain how feedback systems could develop on their own or how they would know (information) what the optimal levels would be. Take, for instance, the regulation of body temperature. How would undirected chance know that the body needs to maintain a certain temperature? How would it know what the optimal temperature should be? Natural selection would not be of any help. There must be something (or someone... a Creator, for instance) that originally designed the entire package. He knew the body's optimal temperature had to be near 98.6. There is nothing in nature that could stipulate this on its own. Natural selection would be of no help because, for the body to live, all of these feedback systems and regulators have to exist at once. If the body cannot maintain homeostasis in all of its systems, it would soon overheat, dry up, or fail in any one of the multitude of critical functions it must have to continue unabated.

Why do we sweat? When you exert yourself to a certain point, you begin to sweat. This is homeostasis at work, attempting to maintain temperature. Signals are sent to the appropriate locations, and water is released while the skin is cooled as the water evaporates. When homeostasis is reached, the sweating stops. How does it know to stop sweating? If we did not stop sweating, we would dehydrate and die. That is why it is important to rehydrate after you have been sweating. Unfortunately, it is not only water that you lose but other things as well. Undirected chance cannot account for sweating. Natural selection would never have time to work because the entity would die before it could select the trait. There are a multitude of these systems in the body, working to maintain your health. It is when these systems break down that trouble begins.

The diagram (figure 1a) is an oversimplified example of how the body maintains homeostasis. It is a program that runs to keep things in balance. The body has a thermostat that will turn on the heat when it gets cold and turn on the air conditioner when it gets too hot. Temperature is just one of the multitude of systems that must be controlled.

Armed with "**INFORMATION**," a program could be designed that would maintain homeostasis. However, knowledge of all the working parts would be required for it to function properly. Changing one aspect of your body's system without knowing how it will affect other parts is a recipe for disaster. Therefore, this is an all-or-nothing proposition. It requires a designed operating system that is aware of all its parts and how they interrelate.

Why is body temperature so critical? When the body reaches 107.6 degrees Fahrenheit, the enzymes that keep the body operating are destroyed, and the body dies. If the body falls below 77 degrees Fahrenheit, the enzymes slow their process, and the body eventually dies. We have a range of about 30 degrees in which the body can survive. However, people who have reached a temperature lower than 77 degrees have survived for a certain amount of time, such as those who fall into an ice-covered lake; for instance, life cannot be maintained very long, even in that situation.

When the body gets cold, it stops sending blood to the extremities and concentrates on internal organs and the brain. How does it know to do that? You then involuntarily begin to shiver, which generates some heat. Certainly, blind chance would not know that; no, it was designed by our creator.

Nervous System

The nervous system does not store nutrients, so it must receive its supply continuously. If that supply is interrupted, it can result in brain damage and/or death. The nervous system maintains homeostasis by controlling and regulating parts of the body. When the body gets out of tolerance, the affected receptor sends a signal through the nervous system to the regulatory section in the brain, and the process begins. In the volume on the nervous system, we will go into greater detail.

Water homeostasis.

Water moves into a cell through osmosis. This is a passive movement not requiring ATP energy to enter the cell. If there is an excess amount of water in the blood, the cells will swell; conversely, if there is too little, the cells will shrink. This is regulated by the hypothalamus. When the hypothalamus recognizes that action is needed, it will release a hormone known as ADH (Anti-Diuretic Hormone) to the pituitary gland. The pituitary then releases the hormone into the bloodstream. ADH acts on the kidneys, and more water is released into the blood from the filtrate.

In the kidney, the Loop of Henle (named after the discoverer) is a process that filters out the salt and maintains a level of water. When there is a large amount of salt in the blood, more ADH will be released, allowing water to flow through the kidneys' collecting tubes. Part of the Body Map series will cover, in more detail, the function of various organs.

Dr. Donald Johnson, in his Science Integrity Series, asks some excellent questions (from "Programming of Life" - www.djpol.info). Science needs to provide plausible mechanisms to explain how nature:

> formally solves life's other complex problems and writes the programs?
>
> develops the operating systems and programming languages?
>
> develops the arbitrary protocols for communication and coordination among the thousands (or millions) of computers in each cell?
>
> develops an alternative generation of prescriptive messages using techniques such as overlapping genes, messages within messages, multi-level encryption, and consolidation of dispersed messages?
>
> defies computer science principles by avoiding software engineering's top-down approach required for complex programming systems?
>
> produces complex functional programs without planning by randomly modifying existing algorithms?
>
> simultaneously modifies multiple such programs to result in the production of irreducibly complex structures?
>
> writes the prescriptive programs needed to organize life's metabolism?

It is like building a factory that will turn out an automobile. The blueprint, with all of the specifications, must be in place for the model of car you are building if it is to fit together and operate properly. It takes careful planning and engineering. Every part must be manufactured to precise specifications. If the frame is too small for the engine, if the fenders are too long, if the drive shaft is not right, or any number of other parts are not designed and manufactured precisely, it will not work. It would be insane to believe that a whole factory could put itself together without a coordinated plan. Yes, we have a blueprint; it is called DNA. But DNA cannot build anything without the program (the intelligence to operate it). It is not just the physical parts, but the feedback systems necessary to keep it going to maintain homeostasis. We are not talking about a Model T here; the human body is a million times more complex than an automobile.

To imply that undirected chance can account for the organization and programming of this complex structure, called the Homo sapiens, is beyond ludicrous! It takes knowledge, information, and a designer. All of the parts must work together, or the system fails. I have taken university-level biology courses and have read their material; one thing is common: they all bow to the god of evolution without batting an eye, stating as fact something that has to be taken strictly by faith. It is no wonder so many lose their faith in God when attending these godless indoctrination centers called universities.

Without that knowledge, these systems could never be developed, and certainly NOT by chance! The Creator (the Great Designer) was aware of all systems and functions of the body. Some systems are dependent on others for proper operation. This reminds me of programming. If a system is to operate effectively, you must know, before you start, how to deal with events that are out of predetermined tolerances. You also must be aware of other parts of the system. If you do not consider the entire program, you may write some code that has an adverse effect on the other parts. This is why complete system testing must be done whenever changes in code are added. I have seen many times where a fix to one section of code makes another section of code go bad.

When the programmer (designer) is aware of the specifications (information), they can program responses to an issue that may or may not arise (knowledge). The operative words here are knowledge and information. Undirected chance has neither.

For instance, if the pancreas is not operating correctly, blood sugar gets out of control, and a cascade of ill effects begins to occur. When homeostasis is not maintained, ill health follows.

Failure in Homeostasis: Diabetes

The pancreas is what controls blood sugar. When it is too low, the pancreas releases glucose to maintain homeostasis. When it is too high, the liver synthesizes and stores it. Glucose homeostasis is governed by the interplay of insulin, glucagon, amylin, and incretin hormones. We are all familiar with insulin and glucagon. Insulin is sent out from the pancreas to tell cells to accept glucose from the bloodstream. If blood sugar is still too high, insulin will bind to the liver and cause it to take in glucose. Glucose is a monomer, meaning a single unit. The glucose molecules come together in the liver to form glycogen, a polymer made up of several glucose molecules.

When blood sugar levels fall, the pancreas excretes the hormone glucagon, which binds to the liver, and the liver converts the stored glycogen molecules back into glucose and releases it into the bloodstream.

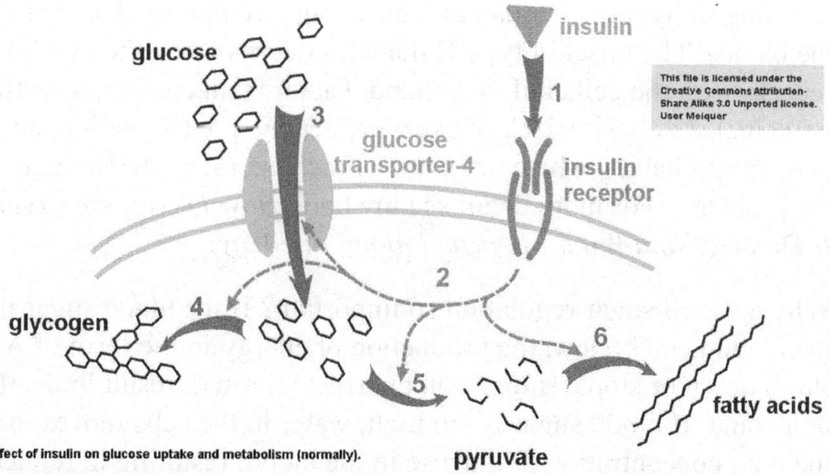

Effect of insulin on glucose uptake and metabolism (normally).

1. Insulin binds to its receptor
2. which in turn starts many protein activation cascades
3. These include: translocation of Glut-4 transporter to the plasma membrane and influx of glucose
4. glycogen synthesis
5. glycolysis
6. and fatty acid synthesis

Amylin was first discovered in 1987 and was found to help the regulation of glycemic levels in the blood. Amylin is released with insulin in the b-cells. Here is how amylin works: as stated in the article "*Amylin replacement with pramlintide as an adjunct to insulin therapy improves long-term glycaemic and weight control in Type 1 diabetes mellitus: a 1-year, randomized controlled trial*". The overall effect is to slow the rate at which glucose appears in the blood after eating. This is achieved by a coordinated set of physiological actions: slowing gastric emptying, inhibiting digestive secretions such as gastric acid, pancreatic enzymes, and bile ejection, and ultimately reducing food intake. These mechanisms were detailed in a study by Ratner et al. (2004), which demonstrated that amylin replacement with pramlintide, when used as an adjunct to insulin therapy, significantly improves long-term glycemic control and supports weight management in individuals with Type 1 diabetes mellitus over the course of a one-year randomized controlled trial.

In the case of type II diabetes (90% of cases), the issue can be a result of many factors: the pancreas does not create enough b-cells (which are used to store and release insulin), and the insulin receptors on the cells become insensitive to insulin signaling, resulting in hyperglycemia (an increase in circulating glucose in the blood). The onset of type II diabetes occurs when the insulin receptors on the cells fail to respond. Factors that contribute to this issue include obesity, lack of exercise, genetic predisposition, and poor eating habits, which are some of the general contributors to the problem. (For more detail see my book *How I Overcame Type II Diabetes and What I Learned Along The Way*)

Why is blood sugar regulation so important? If the blood sugar in the blood gets too low, the production of energy in the form of ATP slows down or stops. If this is not corrected, it will result in death or a coma. If blood sugar is too high, water in the cells moves to the high concentration of glucose in the blood, resulting in reduced water in the cells. If the blood glucose level is not reduced, you become thirsty, or in extreme cases, dehydrated. Regular sodas are overly concentrated with sugar; since this raises the sugar levels in the blood, water is extracted from the cells. This is why sodas never quench your thirst but make you thirstier. The tendency is then to drink more—good for the soda manufacturers, but not so good for us.

Our bodies were designed to operate in a specific way. If we overwhelm them with overly processed foods and chemicals (see endocrine disruptors later in this chapter) and a lack of movement, the result is poor health. To blame God for death and disease is to point the finger in the wrong direction. It is sin, greed, and a lack of human understanding that causes most of our issues, just like the soda companies that put enormous amounts of sodium and sugar in their products to increase thirst and the need to drink another "Mega Soda."

Inslin Resistance

When I consider insulin resistance, I see it as a clear example of the degradation and unravelling of biological systems from their original, perfect design, a consequence of the Fall. Our bodies, intricately designed by a wise Creator, were initially programmed for optimal health, efficiency, and longevity in a perfect world. Insulin resistance, therefore, isn't a design feature, but rather a manifestation of a fallen world's impact on a beautifully engineered system.

1. The Original Design: Intricate and Efficient Metabolic Control

Before the Fall, God created humans with perfect metabolic systems. Insulin, a key hormone, was designed to be a highly efficient messenger, precisely regulating blood glucose levels and nutrient partitioning. Our cells were perfectly attuned to insulin's signals, ensuring that energy was stored and utilized exactly as needed. This efficient system would have prevented the kind of chronic energy surplus and cellular overload that underlies insulin resistance today.

<u>Perfect Cellular Communication</u>: In the unfallen state, cell membranes would have been pristine, insulin receptors perfectly

configured, and intracellular signaling pathways flawless. Every cell would have responded precisely to insulin's call to take up glucose, amino acids, and fats, maintaining optimal energy balance.

Ideal Environment: The original creation provided a perfect environment with nutrient-dense foods, appropriate physical activity, and stress-free living. These factors would have naturally supported healthy insulin sensitivity.

2. The Impact of the Fall: Degradation and Dysregulation

The Curse, as described in Genesis 3, introduced decay, disease, and death into creation. From a creation science perspective, insulin resistance is one of the many biological defects that arose as a result of this spiritual and physical degeneration.

Genetic Entropy: The human genome, though originally perfect, has been accumulating mutations over generations. While not always directly causative, these genetic changes can predispose individuals to impaired insulin signaling or a diminished capacity for cellular repair, making them more susceptible to insulin resistance when confronted with modern environmental factors. Environmental Degradation: The fallen world now presents challenges that our original design wasn't meant to withstand. This includes:

Processed Foods and "Nutrient-Poor" Diets: The abundance of refined carbohydrates, unhealthy fats, and artificial ingredients in modern diets constantly floods the body with excess energy and inflammatory compounds. This overwhelms the finely tuned metabolic system, forcing the pancreas to overproduce insulin. Sedentary Lifestyles: God designed us for activity. Modern sedentary habits reduce muscle insulin sensitivity, as muscles are no longer regularly engaging their glucose uptake mechanisms through physical work.

Stress and Sleep Deprivation: Chronic stress elevates cortisol, which can counteract insulin's effects. Poor sleep further disrupts hormonal balance, contributing to insulin resistance. These stressors are amplified in a fallen world.

Toxicity: Exposure to various environmental toxins can interfere with cellular function and contribute to inflammation, indirectly impacting insulin sensitivity.

3. Insulin Resistance as a "Compromise" in a Broken System
From our perspective, insulin resistance can be seen as the body's attempt to cope with an unhealthy environment, but at a cost. When cells are constantly bombarded with excess nutrients, they develop resistance to insulin's signal to prevent further overloading. This is a survival mechanism, but one that ultimately leads to chronic disease. The body is trying to protect itself from "energy toxicity," as some researchers put it, but in doing so, it sets the stage for a cascade of problems.

4. Hope and Responsibility: Mitigating the Effects of the Fall

Despite the reality of the Fall, creation science emphasizes God's continuing common grace and the inherent design principles that still operate. This provides a framework for understanding and addressing insulin resistance:

Understanding the Design: By studying the intricate mechanisms of insulin action, we can appreciate the Creator's wisdom. This knowledge guides us toward interventions that align with the original design.

Lifestyle Interventions as Restoration: Dietary changes emphasizing whole, unprocessed foods (closer to what might have been consumed pre-Fall), regular physical activity (utilizing the body as designed), adequate sleep, and stress management are not just "treatments" but rather attempts to bring our bodies back into alignment with their intended function. They are acts of stewardship over the bodies God has given us.

For us, insulin resistance is a poignant reminder of a world groaning under the weight of the Fall. It highlights how deviations from God's original, perfect design for our bodies and our environment lead to dysfunction and disease. Yet, it also underscores the enduring wisdom of the Creator, whose design principles offer the best path to health and well-being, even in a fallen world.

Calcium Homeostasis

There are two hormones that regulate calcium homeostasis: parathyroid hormone and calcitonin. When calcium is low (where did the information come from that determines what is an optimal level of calcium?), the parathyroid gland sends a hormonal signal to the bones to release calcium into the blood to be used by cells in the body. In cases where calcium is perceived to be too high, calcitonin is released to slow the release of calcium. Calcitonin is produced by the C-cells in the thyroid gland.

When too much calcium has been released, a condition known as hypercalcemia occurs. This condition, if not corrected by calcitonin, can cause kidney stones, organ damage, and osteoporosis because too much calcium is being released from the bones.

Hypocalcemia is caused when there is not enough calcium in the blood and cells. This condition can also cause osteoporosis, muscle spasms, and osteomalacia, which is the softening of bones. In children, it is known as rickets. Because of the lack of vitamin D and poor absorption of calcium, this condition can cause severe deformation. In adults, it causes muscle weakness, various body pains, and fragile bones.

Look at an airplane and marvel at its design and function; then turn around and look at the magnificent design of the human body. Just comparing those two, no one would believe an airplane was accidentally formed, and yet today's evolutionary scientists would have us believe in evolution. This is insane! I do not believe that even Darwin, if he had known the things that we now know, would have dared suggest that accidental blind chance could ever have structured what we see today. Even so, he did express concern about how sight could be developed with slight modifications.

But Satan knew the best time to unleash the lie: a time when scientific discovery was becoming an everyday occurrence, where speculation could run wild in the minds of men with very little actual understanding. The age of reason was taking over religious belief. Because of all these conditions, science began to become a religion based on faith in blind chance, called evolution.

We can excuse the early proponents of evolution for their lack of understanding. We cannot excuse the willingly blind scientists today. They know better but refuse to acknowledge the Creator. God is giving them over to a reprobate mind. Look at the completely insane ideas that they come up with to avoid accepting the truth that all things were created in the beginning by Him. In Appendix B, we point out just a few of the unsubstantiated claims of the evolutionists. I encourage you to be aware of these examples and not be taken in by the fraud.

Blood clotting - An amazing feat of design and programming

Although not technically homeostasis, blood clotting is a very critical factor in maintaining life. Have you ever stopped to think about how you stop bleeding after a minor cut? If the body did not have a mechanism for stopping the blood from flowing, you would die in a fairly short time. If the coagulation process did not stop at the wound site, it might coagulate all the blood in your body once the process started, and you would die within minutes. However, the Great Designer has created a fascinating system for handling this problem. Blood clotting is a multi-step process with many dependencies. The first step is vascular constriction, limiting blood flow to the injured area. Next, platelets become activated by thrombin, a protein found in the bloodstream. This forms a temporary platelet plug. The protein fibrinogen is responsible for platelet clumping. The platelets clump by binding to collagen, which is exposed by the cut in the lining of the vessels. When activated, the platelets release ADP and thromboxane A2, a lipid known as an eicosanoid. Both of these release additional platelets, serotonin, phospholipids, lipoproteins, and other proteins critical to the coagulation cascade. The platelets then change their shape to help form the plug. Next, the platelets are meshed in a network of fibers known as a fibrin mesh, and the clot is formed. Finally, after a few days, plasmin begins to dissolve the clot, and repair is complete. In actuality, the blood clotting process is much more complex than what we just described. If you look at the diagram on the following page, you will see a high-level diagram of the clotting process. There are two pathways: the intrinsic pathway, which deals with problems within the vascular system, and the extrinsic pathway, which deals with breaks in the vessel wall. There are 12 factors used in the clotting process. The coagulation factors are manufactured in the liver (except factor IV, see the chart on the next page). Factor I - fibrinogen

Factor II - prothrombin

Factor III - tissue thromboplastin (tissue factor)

Factor IV - ionized calcium (Ca++)

Factor V - labile factor or proaccelerin

Factor VII - stable factor or proconvertin

Factor VIII - antihemophilic factor

Factor IX - plasma thromboplastin component, Christmas factor
Factor X - Stuart-Prower factor

Factor XI - plasma thromboplastin antecedent

Factor XII - Hageman factor

Factor XIII - fibrin-stabilizing factor You can follow the extrinsic process on the diagram. In the case of a cut on the outside of the body, the tissue factor and factor VIIa (active form) activate factor X, forming factor Xa. Factor Xa is then able to activate prothrombin (II) to form thrombin (IIa). Thrombin then converts fibrinogen to fibrin (factors I and Ia, respectively). Fibrin forms a mesh that, in conjunction with the platelets, plugs the break in the vessel wall. The fibrin mesh is further stabilized by factor XIII.

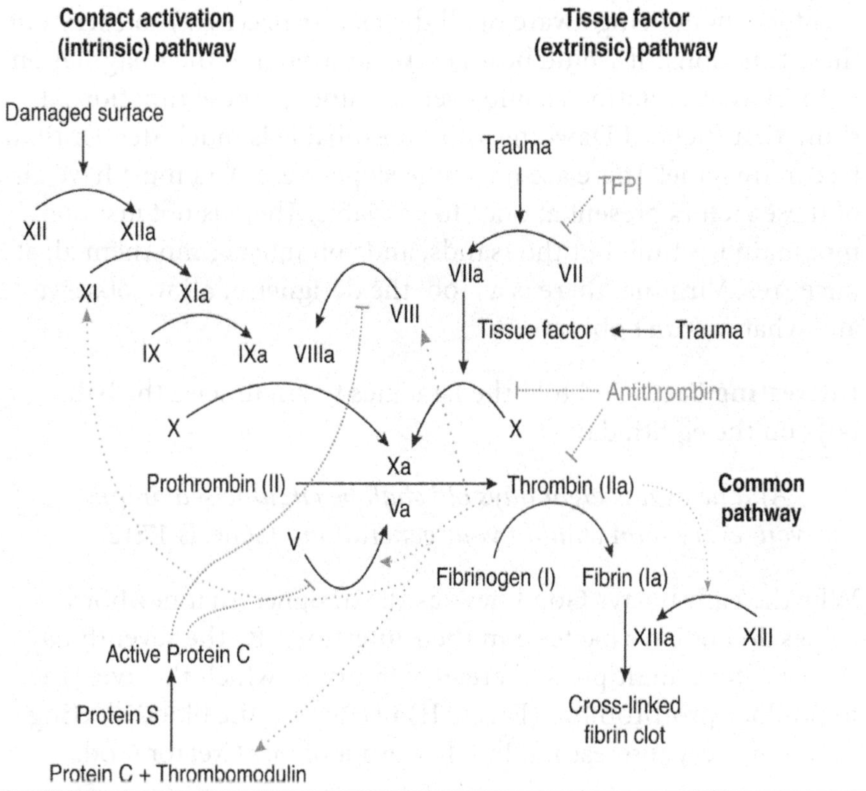

Problems with clotting arise when a defective gene is passed to a newborn; this is known as hemophilia. Hemophilia lowers blood plasma clotting factor levels. The missing coagulation factors prevent fibrin formation, so the wound never completely clots. This is particularly dangerous with internal bleeding. Once again, we see that mutations are not helpful to the evolutionary cause.

The human body is made up of many complex interrelated systems. These systems can communicate and somehow initiate the processes automatically to protect the body from significant irreversible damage.

We will look at the immune system and the process of blood coagulation in detail in another volume. Each function is a combination of complex processes, without which the human body could not survive. These are only two of the many interrelated systems in the body. Could these arise from undirected chance events without being aware of all the factors necessary to carry out these functions? It would be an extreme stretch of the imagination to think that mutations could ever account for these functions. I think that Richard Dawkins' Mount Probable is much steeper than he can imagine! There are no gentle slopes here. You must have all of these factors present at once to be viable. There is not just one mountain to climb but thousands, and you must climb them all at once. Yes, Virginia, there is a God, the designer of all we observe and what we can't observe.

Interesting Fact: God told the Israelites to circumcise the baby boys on the eighth day.

> *"And he that is eight days old shall be circumcised among you, every man child in your generations* Genesis 17:12.

Why the eighth day? God knew, as the designer, that newborn babies do not have bacteria in their intestines. By the seventh day, these bacteria multiply and create vitamin K, which the liver uses to produce prothrombin (Factor II) needed for the blood clotting function. Very interesting! Further proof of our Creator God.

Acid-base homeostasis (pH): potential hydrogen or the concentration of hydrogen ions in the blood. It is measured logarithmically and reveals how acidic or alkaline an aqueous solution is. The measurement of pH is logarithmic; 6 is ten times more acidic than 7. What you eat affects your pH level; meat, eggs, dairy, beans, nuts, and seeds are more acidic, whereas fruits and vegetables are typically more alkaline. You normally have a pH of about 7.4 in the blood. You can live when pH reaches a level as high as 7.8 or as low as 7.0; beyond that, enzymes and electrical systems don't operate correctly, and eventually, the body will shut down. The more acidic, "the hotter" things are, the faster they respond. The more alkaline the body, "the colder" things get, the slower they react. When the body cools, electrical reactions become more resistant, and nerve impulses slow down. Enzymes and proteins have an optimal pH range; once outside of it, they become inactive or denatured.

We recognize the remarkable precision and complexity of the human body as evidence of purposeful design. One such example is the body's ability to maintain blood pH at approximately 7.4, a slightly alkaline level that is critical for sustaining life. This delicate balance is maintained through a combination of three primary buffer systems, all working together as if finely tuned by an intelligent Creator to ensure the body functions properly under various conditions.

1. The Bicarbonate Buffer System

This is the most important buffer system in the blood. It consists of carbonic acid (H_2CO_3) and bicarbonate ion (HCO_3). When the blood becomes too acidic (low pH), bicarbonate acts as a base and binds with free hydrogen ions (H) to form carbonic acid, which is then broken down into water (H_2O) and carbon dioxide (CO_2), the latter of which is exhaled through the lungs. Conversely, if the blood becomes too basic (high pH), carbonic acid can release H ions to restore the balance.

Equation:

$$CO_2 + H_2O \rightleftharpoons H_2CO_3 \rightleftharpoons H + HCO_3$$

This reversible reaction allows for rapid response to pH changes and is tightly linked to respiration—another sign of interdependent, complex systems.

2. The Respiratory System

The lungs assist in regulating pH by controlling the amount of CO_2 in the blood. Since CO_2 reacts with water to form carbonic acid, changes in breathing rate can directly affect blood pH. For example, hyperventilation removes CO_2 quickly, raising pH (alkalosis), while hypoventilation retains CO_2, lowering pH (acidosis). The brainstem monitors blood pH and adjusts breathing automatically, maintaining stability—a beautiful example of feedback control.

3. The Renal System (Kidneys)
The kidneys provide long-term pH regulation by excreting hydrogen ions (H) and reabsorbing bicarbonate (HCO_3) as needed. This process is slower than the respiratory response but essential for restoring pH over hours to days. The kidneys can even generate new bicarbonate ions when needed.

A Designed System of Checks and Balances
Together, these systems operate in unison to keep blood pH within a narrow range of 7.35 to 7.45. Any sustained deviation outside this range can be life-threatening. Such precision, complexity, and interdependence argue strongly against random chance and instead point to the intentional engineering of life.

In this system, we see not only order but also redundancy and adaptability, which provide stability under stress. From a creation science perspective, this is not merely chemistry—it is evidence of a Creator who designed the body with wisdom and foresight, equipping it to thrive in a dynamic and often challenging world.

Exercise

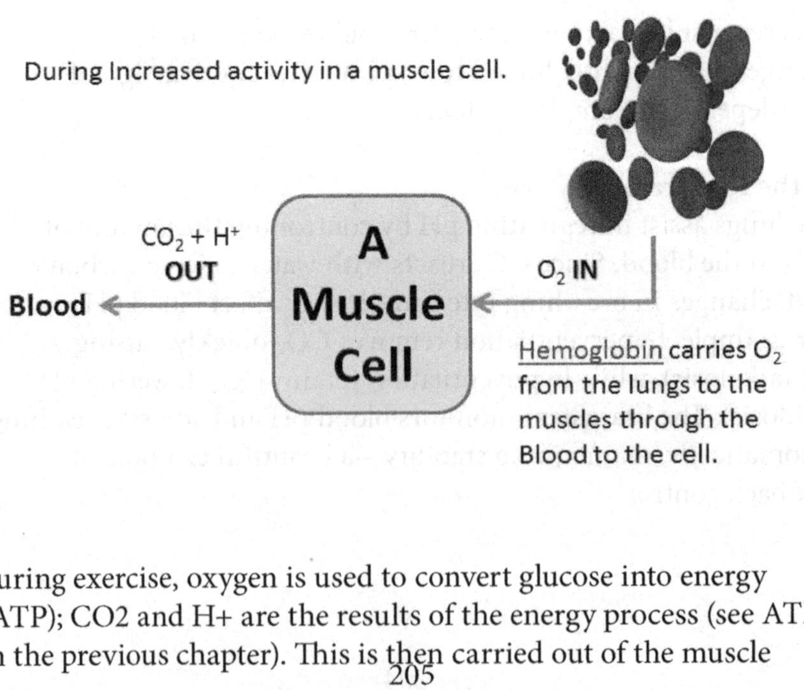

during exercise, oxygen is used to convert glucose into energy (ATP); CO2 and H+ are the results of the energy process (see ATP in the previous chapter). This is then carried out of the muscle by the blood; this causes the pH level to drop. Because of the wonderful design of our bodies, we have buffers. The kidneys will remove excess chemicals from the blood; if the kidneys cannot remove the chemicals fast enough, the result is metabolic acidosis. This is where the lungs play their part in removing excess CO2 through heavy breathing. If the lungs cannot remove enough CO2, the result is respiratory acidosis.

Different parts of the body have different pH levels depending on their function. Gastric acid is generally around 1, so the digestive system can "digest" foods. The stomach's pH is between 4 and 6.5; the lower part of the stomach is very acidic, between 1.5 and 4. When the food enters the small intestine, the pH is more alkaline, with a pH of 7 to 8.5. This is where the absorption of nutrients occurs; the waste product is sent to the colon, which is again very acidic, with a pH of 4 to 7. If your body is not slightly alkaline, the body cannot heal itself. If your pH is not balanced, you will not be able to assimilate nutrients effectively.

If your body is not slightly alkaline the body cannot heal itself. If your pH is not balanced you will not be able to assimilate nutrients effectively.

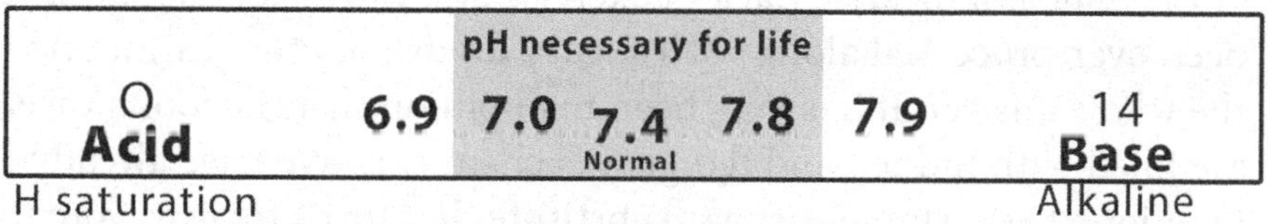

Blood Pressure Homeostasis:

Low blood pressure is called hypotension and can be caused by things like dehydration and loss of blood from a major cut. Hypotension, if not corrected, can cause fainting or, worse, kidney failure. Hypertension occurs when blood pressure is too high and can ultimately, if not brought into balance, cause a stroke.

There are two hormones that affect the cardiovascular system: ADH (anti-diuretic hormone - stimulates contraction of arteries and capillaries) and epinephrine (known as adrenaline). The baroreceptor (baro means pressure) found in the carotid sinuses (blood that is going to the brain) and the aortic arch (blood going to the body) are receptors. These receptors communicate changes in the stretch of the veins to the brain. The brain remembers what your normal set point is, say 120/80, because of the number of action potentials it receives through the central nervous system. If that number goes up, the brain knows that blood pressure is above normal. If the number of action potentials goes down, then the brain knows that blood pressure is dropping.

There are two inputs from the nervous system to the heart: sympathetic input and parasympathetic input. Sympathetic input will increase the heart rate, pumping more blood, while parasympathetic input will decrease the rate.

Additionally, endothelial receptors control the constriction of blood vessels. These receptors are found in the smooth muscle tissue of blood vessels. Endothelins attempt to maintain a balance between vasoconstriction (narrowing) and vasodilation (widening) of blood vessels.

Endocrine Disruptors

I have often wondered why there is a great increase in obesity and diabetes, not only in adults but also in children! I am sure that part of the issue is a lifestyle (lack of exercise and eating foods that have been over-processed along with sugary foods), but that cannot be the whole answer. It has long been my opinion that the foods we eat are filled with hidden and dangerous chemicals. We have already mentioned aspartame, a sugar substitute that turns toxic in your body, but are there other hidden chemicals that destroy the natural functioning of the body?

Are food companies knowingly or unknowingly making you fat and diabetic? A new study by the Endocrine Society says yes. In the meeting of the International Conference on Chemicals Management (ICCM4) in Geneva, Switzerland, the Endocrine Society stressed the importance of reducing EDCs (endocrine-disrupting chemicals) found in our food.

We have already illustrated how important hormones are in maintaining good health. EDCs work by interfering with the body's natural hormones and therefore disrupt the ways cells develop and grow. One of the more well-known EDCs is bisphenol A (BPA). This chemical is commonly found in plastics, cosmetics, food can linings, and many other common everyday products. According to the Society, virtually everyone on earth has most likely been exposed to some degree. About 35 percent of American adults are obese, and more than 29 million Americans have diabetes, according to the Society's Endocrine Facts and Figures report.

> "Hundreds of studies are pointing to the same conclusion, whether they are long-term epidemiological studies in human, basic research in animals and cells, or research into groups of people with known occupational exposure to specific chemicals." - *Andrea C. Gore, Professor and Vacek Chair of Pharmacology at the University of Texas at Austin* - *[Http://medicalxpress.com/news/2015-09-chemical-exposure-linked-diabetes-obesity.html]*

In the paper titled "***Environmental Obesogens: Organotins and Endocrine Disruption via Nuclear Receptor Signaling,***" authored by Felix Grun and Bruce Blumberg from the Department of Developmental and Cell Biology at the University of California, Irvine, it is suggested that while increased calorie consumption and insufficient physical activity are significant contributors to obesity and diabetes, these factors may not encompass the full scope of the issue.

Recent research has brought attention to the potential role of environmental obesogens—xenobiotic substances that can interfere with the typical processes regulating fat development and energy homeostasis. Environmental estrogens, which are chemicals possessing estrogen-like properties, have been found to disrupt adipogenic processes in laboratory settings. Additionally, other categories of endocrine-disrupting chemicals (EDCs) are now being examined for their effects.

The authors argue that obesogens, substances that affect lipid regulation and fat storage, can alter energy equilibrium or change appetite regulation, thereby encouraging fat accumulation and, consequently, obesity. These xenobiotic agents—compounds not naturally present in our food, including various drugs and environmental chemicals—can disturb the standard regulatory mechanisms governing fat cells and energy balance. Certain chemicals exhibit estrogenic properties, typically associated with female hormones, which can interfere with normal cellular processes related to energy storage. Current investigations are focusing on EDCs as potential contributors to the development of obesity and the emergence of diabetes.

Their study also indicates that the child is more likely to be obese if the mother is a smoker before or during pregnancy but not necessarily after the birth of the child, leaving us to assume that chemicals from smoking cross the placenta barrier and affect the development of the child. Studies have shown that there are at least 20 obesogens commonly found in our environment. The most widely used obesogens are BPAs, phthalates, and perfluorooctanoic acid (PFOA), and to a lesser degree, tributyltin.

Bisphenol A (BPA) is commonly found in water bottles and the lining of metal cans, probably the most widely known EDC. In the 1930s, it was used in animal feed as a synthetic estrogen to stimulate growth.

Phthalates are found in toys, pharmaceutical products, food containers, paints, and many other consumer goods. Phthalates interrupt hormone receptors that regulate lipid and carbohydrate metabolism. This chemical is associated with an increase in abdominal obesity and weight gain.

Perfluorooctanoic acid (PFOA) is commonly associated with Teflon cookware. PFOA is linked to thyroid disease, chronic kidney disease, and cancer.

Tributyltin (TBT) is an organotin. At the atomic level, organotins are compounds consisting of one to four organic components attached to a tin atom via carbon-tin covalent bonds. According to the website Http://toxipedia.org/display/toxipedia/Tributyltin:

The primary application of organotins lies in their role as stabilizers during the production of plastic goods, serving functions such as preventing yellowing in transparent plastics and acting as catalysts in poly(vinyl chloride) products. Additionally, organotins have a more limited use as biocides, which include fungicides, bactericides, and insecticides, as well as preservatives for materials like wood, textiles, paper, leather, and electrical devices.

In the case of TBT, it is altering lipid homeostasis, causing fat cells to grow or influencing the MSCs (stem cells or preadipocytes - see stem cells in the chapter on the not so simple cell) to become fat cells (adipocytes), thereby increasing the number of fat cells by altering gene expression."

Hormones affected by EDCs

The hormones insulin and glucagon play a crucial role in controlling blood glucose levels. Ghrelin, another hormone, triggers feelings of hunger, while leptin and adiponectin, which are secreted by fat cells, help to regulate appetite. Low leptin levels are linked to obesity, similarly, elevated adiponectin levels can also contribute to the same outcome. Furthermore, the thyroid gland influences the metabolic rate during periods of rest. According to the obesogen hypothesis, "chemicals are a contributing factor to obesity." (excerpts from Bruce Blumberg, Professor of Developmental Biology, University of California, Irvine).

How?

1. Creating new fat cells (adipogenesis).

2. Increasing the fat we already have by increasing appetite, decreasing satiety (gratified beyond the point of satisfaction), and

altering the metabolic rate (the rate at which fat is used for energy).

The EDCs work by mimicking a natural hormone, and the result is over-stimulation of a natural process.

EDCs (endocrine-disrupting chemicals) affect us in ways we are unaware of at the cell level. Some of the effects of EDCs do not show up until adulthood. We discuss the endocrine system in more detail in Book Six.

What does the endocrine system regulate?

Development and growth, sexual maturation, reproduction, brain function, thyroid function, metabolism, immune functions, and cardiovascular functions.

The endocrine system works with hormones that communicate with cells and cause specific actions to be taken. The Endocrine Society is suggesting that something be done to restrict the use of these EDC chemicals. Anyone who is obese or diabetic should research this further to determine if their lifestyle includes these chemicals and what can be done about it.

Endocrine Hormones

Hormones are chemical messages that move through the body in the blood. Only those cells that have the specific receptors will receive the signal meant for that cell type. The following is a list of endocrine glands and their hormones and the systems affected.

Hypothalamus: Excretes hypothalamic-releasing and release-inhibiting hormones. It regulates the anterior pituitary hormones. The hypothalamus controls: body temperature, hunger, thirst, fatigue, sleep and Circadian rhythms.

Adrenal cortex

Glucocorticoids (e.g. cortisol) promote gluconeogenesis.

Mineralocorticoids (e.g. aldosterone) promote sodium reabsorption by the kidneys.

Homeostasis is critical for good health and life. We have discussed several different systems for keeping the body in balance; each one has a critical function. Once again, we have an all-or-nothing scenario; these systems exist from the beginning, or the organism would die.

By now, you should be getting an idea of how complex the body is, with all systems working in concert. It is like a symphony orchestra; if you handed different scores to different instruments, the sound would be nothing but noise. However, with all of the instruments playing the same score, you have some of the most beautiful music in the world. Our bodies were designed from the beginning by the Great Composer; only intelligence could design such a magnificent creature.

Tissue repair

Not technically referred to as homeostasis, it is a process that keeps the body in repair. Without tissue repair, we would have a mass of scars that never heal, and over time it would not be a pretty sight. Unfortunately, many major cuts and burns do not heal and create permanent scars. There are four stages:

1. Blood clotting (homeostasis) (see blood clotting earlier in the chapter)

2. Inflammation - at this time, damaged and dead cells are removed from the site by white blood cells in a process called phagocytosis.

3. The growth of new tissue (proliferation). Platelet-derived growth factors cause cells to speed up their division and growth.

4. The remodeling of tissue (maturation). We will describe this process in the volume on organs, in which the skin is detailed.

THE IMMUNE SYSTEM
DEFENSE AGAINST DISEASE
T
B
K

6

The Immune System

An introduction to the defense against dark forces

This chapter is an introduction to the immune system. A detailed description of the system will be included in book VIII on the lymphatic system.

Human physiology is an amazing product of a highly intelligent designer, God. How would the trillions of cells ever get so thoroughly coordinated? Small steps of evolution? No, I think not. After conception, what "Life Force Director" is in charge of building a human being from the blueprint (DNA) it is given? How does it know what to build first, second, etc...? Hormones send messages to different parts of the body. How do the glands know when and where to send the hormone? There are messages being sent continuously throughout the body; how is it all communicated? After the early twenties, how does the system know to stop growing? There are many questions like these we could ask, but the evolutionist makes empty statements like "it evolved" or "it adapted," but that does not explain how it originated. This is always the way of the evolutionist: make "matter of fact" statements, but never explain exactly how each successive step caused the thing to evolve in the first place.

We are told that evolution accounts for all living systems, but we are not told exactly how. Tell me, in intricate detail, how the eye could evolve step by step. This is the trick of the evolutionist's mind frame: they will make claims about what may have occurred. With great confidence, they declare evolution as fact. Where is the evidence? Ask an evolutionist how the first life evolved and continued to evolve to ever more complexity, and get ready for a "why" or "what could have happened," but never HOW.

In his book *Climbing Mount Improbable* (W. W. Norton & Company, 1996), Richard Dawkins proposes that the solution to understanding complex biological structures lies in breaking down their improbability into smaller, manageable steps. He suggests that evolution does not scale the steep, seemingly impossible cliff of complexity in a single leap. Instead, it "acts by breaking the improbability up into small manageable parts, smearing out the luck needed, going round the back of Mount Improbable and crawling up the gentle slopes," emphasizing a gradual, accumulative process over time.

To me, this sounds like a lot of psychobabble. He still does not explain how an extremely complex structure like the human eye can evolve on its own, undirected. Where does the new information come from that directs these small steps? This new information must be coded in the DNA of the father's sperm or mother's eggs to be passed on. How does that happen? Mutations? We have already learned from current observations and testing that mutations do not increase information. In fact, we have observed that all mutations are virtually always harmful. Natural selection would not "freeze" these new features unless they were, in some way, beneficial according to their hypothesis. It is still speculation and conjecture, and there is no evidence for evolution. It is wishful thinking on Dawkins' part to hope that these small parts would end up doing anything but causing trouble.

The eye is not just a photoreceptor; it is a complex organ that has to have connections to the brain to work. How did that section of the brain evolve simultaneously? It is a complete system or nothing.

David Berlinski sharply critiques Richard Dawkins' concept of Mount Improbable by pointing out a fundamental flaw in Darwinian reasoning. He argues that the very problem Darwin's theory sought to avoid resurfaces when examining complex systems like the eye. Any beneficial change to one part of the visual system must be matched by coordinated changes throughout the entire system. For example, increasing the number of photoreceptive membranes is meaningless without a corresponding increase in the size and complexity of the optic nerve, which must then be matched by neurological adaptations in the brain. If all these changes occur simultaneously, the idea of a slow, step-by-step climb up Mount Improbable collapses. But if they do not occur together, it's unclear why such isolated, incomplete changes would be favored or even occur at all.

Michael Behe coined the term "irreducible complexity." After careful study of the bacterial flagellum motor, Behe determined that small changes required by evolution could never explain how a very complex structure with multiple parts could evolve. Either all parts were there and functional, or natural selection would have eliminated them.

Geoffrey Simmons, M.D., introduces the term "***whole package phenomenon***" (WPP) to describe the intricate interrelationship of hormones within human physiology. The endocrine system, he explains, is composed of major glands such as the pituitary, thyroid, adrenals, testes, and ovaries, along with numerous smaller hormone-producing tissues located in organs like the pancreas, heart, lungs, kidneys, stomach, liver, and placenta. Each of these glands produces specific hormones that act as messengers, directing target cells on what to do, when, and how. Currently, over 40 hormones are known, each regulated by its own feedback loop. These hormones often work in parallel or tandem; some interact competitively, and others are governed by double or even triple feedback systems. The sheer complexity, the interdependence of all components, and the requirement for all systems to be functional simultaneously highlight what Simmons calls the whole package phenomenon—posing a significant challenge to evolutionary theory by underscoring the necessity of fully integrated systems from the start.

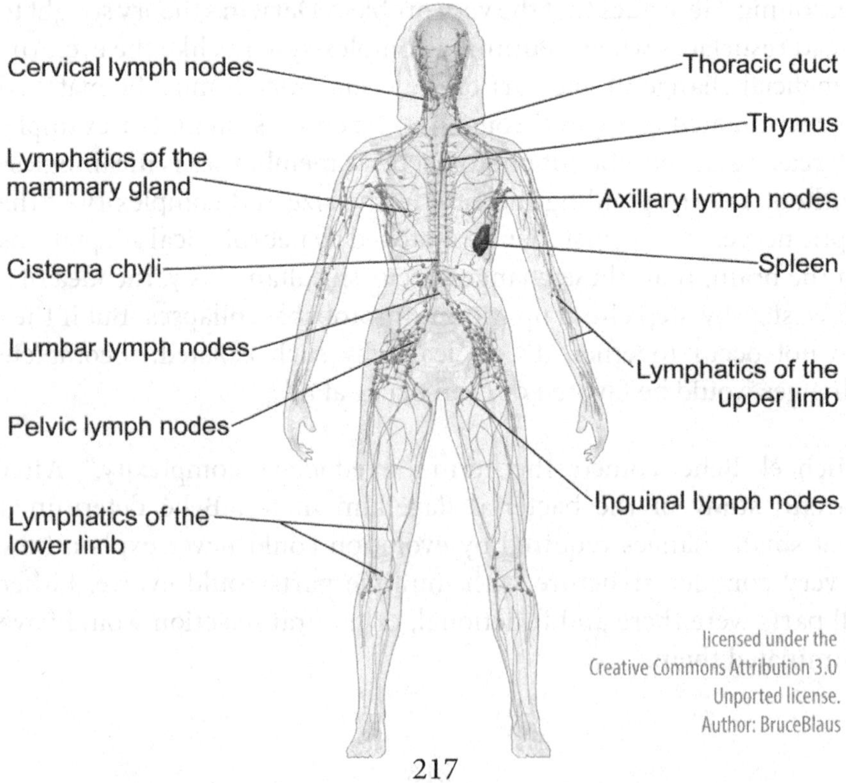

The immune system is another system designed with such amazing precision and complexity that it causes us to stop and praise the Almighty Creator. Again, we have a system that must be fully in place, or the organism will die. There is no way to have a living organism that cannot defend itself against the myriad of invaders that are relentlessly on the attack. You can see from the illustration that the lymphatic system is similar to the bloodstream. However, the lymphatic system does not have the heart to pump liquid lymph but relies on pressure gradients to move its contents. The contents end up back in the bloodstream to be filtered, and waste to be eliminated.

Major Histocompatibility Complex (MHC) Your body has memory; it knows what cells belong to you. MHCs are surface cell identifiers that tell the body which cells are "self." In our bodies, it is also known as the human leukocyte antigen (HLA) system. The "memory database" is created during development and is permanent throughout your life. If a cell does not have the correct MHC, the immune system will try to eliminate it. This is why transplants are so difficult; since the identifiers are not "you," the immune system will try to eliminate the foreign cells. However, if the system malfunctions, it may begin to eliminate cells normally in the body; this is known as an autoimmune disease. There are many types of autoimmune diseases that attack organs, glands, and tissues. This is a serious condition that can lead to death.

There are two or more types of MHCs; type I is found on every cell of the body, and type II is limited to macrophages and lymphocytes, which we will discuss shortly.

The placenta does not have the MHC but is not destroyed by the immune system. We will discuss how that is allowed in book II on reproduction.

The Immune System - Defense against dark forces

The human body is an amazing example of engineering excellence! The human body is an amazing example of engineering excellence! We take for granted what is happening in our bodies each day because we cannot see it happening. We have an immense number of systems working twenty-four hours a day in harmony. These systems work on our behalf without any conscious effort on our part. We breathe, we process and generate energy, we repair, we build, we pump life-giving blood, and we fend off attackers, all without being aware of it.

There are many functions in the human body that must have the cooperation of other systems. One such system is the immune system. Without the immune system, we would not survive long; ask anyone who has Acquired Immunodeficiency Syndrome (AIDS), which is a disease caused by the Human Immunodeficiency Virus (HIV). This disease weakens the immune system, leaving the body without a defense against disease.

The immune system is a system of biological structures and processes that protect the host from pathogens (infectious agents or germs). The immune system detects a wide variety of pathogens but also needs to recognize the body's own healthy cells and tissues to function properly. It is interesting to note that the immune system knows what belongs to your body and what does not. During development in the womb, your operating system is creating a type of database, which is a list of everything that is uniquely you (MHC). The immune system knows, in the case of a transplant, that the new organ is a foreign entity not belonging here and tries to rid the body of it. This is why organ transplants have a tough time getting accepted. I had a friend, John Edwards, who had a liver transplant, and he had to take certain drugs so his body would not reject the new liver. He lived a fairly normal life and was able to live well beyond the statistical lifetime of most liver transplant patients. Unfortunately, he did eventually succumb to the rejection of his liver.

There are two types of immunity: **innate immunity** and **adaptive immunity**. Innate immunity responds quickly to the chemical properties of the antigen based on chemicals in the blood and the immune system cells in the body. Adaptive immunity reacts much slower. The antigen has to be recognized and processed. At this point, the immune system creates a multitude of specific immune cells to fight the new antigen. The body adds this new immune cell to its memory, so if it shows up again, it already has the definition and can respond quicker. The adaptive immune system reaction is why some diseases do not cause the illness a second time, like measles.

An interesting story from history about adaptive immunity comes from our first President, George Washington. When George was 19 years old, he went to Barbados with his brother Lawrence. Lawrence had tuberculosis, and it was suggested that they go to Barbados. Upon arriving in Barbados, they were invited to the home of Gedney Clarke, a family friend and wealthy merchant. George was reluctant to attend because there was smallpox in their home. His fears were realized when he contracted the disease. He was able to overcome smallpox and became well again. (This is the reason he had noticeable scars on his face.)

During the Revolutionary War, there was an outbreak of smallpox. This gave the British the upper hand because they had already survived an outbreak in England and were immune to the disease. It is speculated that the British were involved in a type of chemical warfare by attempting to spread the disease. Washington faced a dilemma: if he inoculated the men, they would be out for some period, giving the British the upper hand; if he did not, he would lose more men to the malady. Washington's decision to inoculate the men proved to be the right one. He could now fight with a full complement of men. According to medical history, smallpox was responsible for more deaths than any other infectious disease. Smallpox was totally eradicated worldwide by 1980; however, the virus still exists under lock and key in Atlanta, Georgia, and Russia. That is very scary since it could be released now, and many in the younger generation would be susceptible to the disease.

There are three basic types of cells in the immune system:

Phagocyte - The word comes from Greek and means "to eat" or "devour." A phagocyte is a white blood cell. Its job is to digest foreign particles, bacteria, and dead or dying cells.

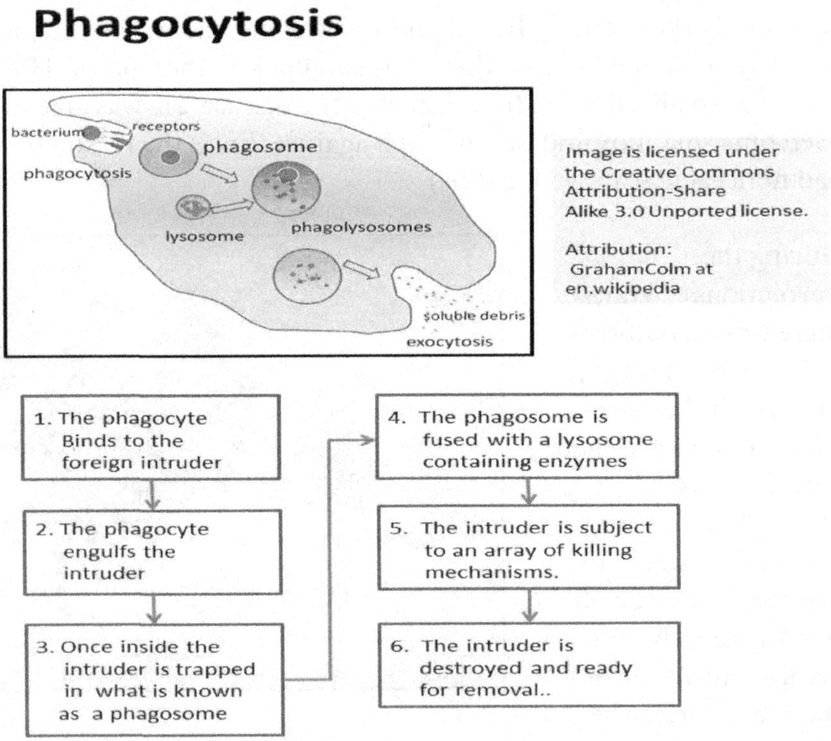

During an infection, the phagocyte is attracted by the chemical signals sent out by the pathogen to the location of the invader. When the phagocyte comes into contact with the invader, the receptors on the phagocyte bind with the invader. The invader is then engulfed by the phagocyte and destroyed with oxidants and nitric oxide. Then, in a process known as antigen presentation, the phagocyte moves part of the ingested material back to its surface where other cells, known as lymphocytes, come into play. Typically, pus is the result of the actions of a phagocyte; it is the leftover of dead cells.

There are three types of lymphocytes: T-cells, B-cells, and Natural Killer cells.

T-cells belong to a group of white blood cells known as lymphocytes. "T" stands for thymus, since this is the principal organ responsible for maturing the cells. However, all T-cells originate from stem cells in the bone marrow. T-cells are involved in cell-mediated immunity functions. In response to an antigen (a substance that prompts the generation of antibodies), the T-cells are activated. Killer T-cells destroy the ingested material that comes to the surface of the phagocyte, as well as the phagocyte cell itself. After the fight is over, most of the T-cells die, except for a contingency of memory T-cells. These memory cells will be called on in case the invader returns. The cell-mediated immunity function is what causes transplanted organs to be rejected.

B-cells also belong to a group of white blood cells known as lymphocytes. These cells play a large role in the humoral immune response. The function of B-cells is to make antibodies against antigens. These work in the bloodstream, fighting anything outside the cell. They eventually develop into memory B-cells after activation by antigen interaction.

There is a two-pronged approach to fighting pathogens: T-cells destroy infected tissue and B-cells kill anything left in the bloodstream. The memory cells will stay with the body for decades, so if the pathogen is encountered again, your immune system is ready to fight. This is the basis used for inoculation.

There are at least three types of molecules involved in the immune response function.

Antigens are foreign substances that cause an immune response and come in many different forms.

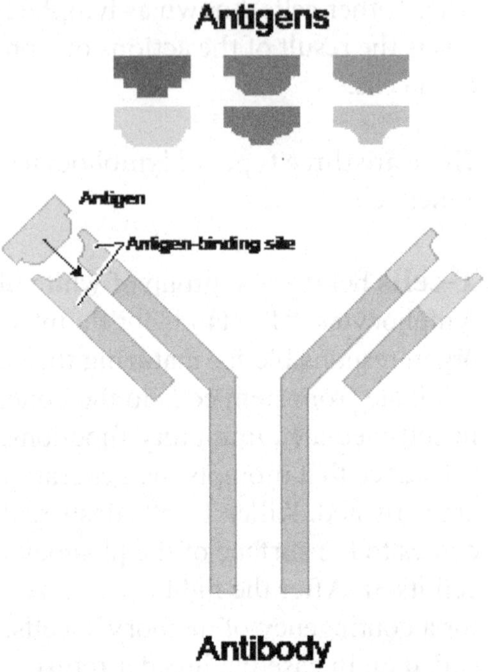

Antibodies identify and neutralize foreign objects, such as bacteria and viruses. I have heard it said that antibodies bind with antigens like a lock and key. The proteins must fit together perfectly for the process to work. There are billions of B-cells in the body; the one chosen to replicate will be the one whose antibodies match the antigen being destroyed. The question you should ask here is: how does it know? Does each of the millions of B-cells check to see if they are a fit? I doubt it. There is intelligence in the process that we do not understand.

T-cell Receptors receptors are receptors on the outside of a T-cell that must bind to the antigen fragments presented by the phagocyte. The T-cell receptors must be joined, just like the antibodies. Again, we can think of it as a lock and key. Both sides must fit perfectly for the process to work properly. There are millions of T-cells. The one chosen to do the work will be the one whose T-cell receptor matches the antigen being presented by the phagocyte. This T-cell is then cloned into an army of specific T-cells to fight the specific antigen.

The whole immune response is an amazing process, but we are not done yet. If we are to fight the millions of potential antigens, we must have millions and millions of different T-cell and B-cell configurations. Unfortunately, there are only 24,000 genes in the human body. How do we get the millions and millions of different protein configurations we need? We have in our DNA separate sub-genes that code for regions of the antibody or T-cell receptor. First, there are heavy chains and light chains of antibodies and T-cell receptors. Additionally, there is a fixed and variable region. Through the genetic manipulation of these variable regions, the antibody and T-cell receptors have the potential for 90 million different proteins, which can be joined to the millions of different antigens. The designer scores again!

Your body is at war 24/7!

You are probably not aware that your body is constantly at war with invaders (antigens). All of the action happens automatically and continuously, so you are not aware of the battles most of the time. If the intruder is known to your immune system, you may not have any symptoms because it can attack the invader very quickly and win the battle without a lot of trouble or inhibit it from entering in the first place. Sometimes, it may take time to attack and win the war, and some results of the battle may appear. In the case of acne, a pus-like protrusion will appear on the face; the immune system has attacked the bacteria on the skin, and the result of the battle is a pimple.

There are three functions of the immune system: to destroy attacking pathogens, to detect and kill aberrant cells, like cancer, and to clear the body of any debris.

First Line of Defense

Your first line of defense is the innate immune system. This type of immunity is referred to as non-specific immunity because it responds to pathogens in a generic way. This system can identify some of the markers on pathogens and eliminate them immediately. Some of the first line defenders are:

Eyes: Lysozyme is found in tears and is antibacterial.

Skin:
Sweating away invaders and keratin, a tough protein on the outer part of the skin. Lysozyme, an enzyme found on the skin that can dissolve pathogens. Desquamation - where dead cells that are sloughed away are replaced.

Mucous:

The digestive and respiratory pathways are lined with it, capturing rogue pathogens with its sticky substance, much like the flypaper we used on the farm in the cattle barn. Then, the cilia sweep the intruder out, where it can be eliminated.

Gastrointestinal tract:
Gastric acid - composed of hydrochloric acid, potassium chloride, and sodium chloride.

Second Line of Defense:

Cellular defenses:

If the antigen enters the body, chemical defenses kick in. These include neutrophils, lymphocytes, monocytes, eosinophils, and basophils.

Leukocytes, or white blood cells, have surface receptors that can detect antigens.

Neutrophils are the most abundant and are the first to respond. They engulf and then destroy the pathogen.

Macrophages are found throughout the body.

Macrophages, derived from the Greek words “makros,” meaning “large,” and “phagein,” meaning “to eat,” are a category of white blood cells that play a crucial role in the immune system. They are tasked with engulfing and breaking down cellular debris, foreign materials, pathogens, cancerous cells, and any substances that do not possess the specific surface proteins characteristic of healthy cells. Functioning as the body's internal sanitation team, these essential immune cells identify and eliminate potentially harmful intruders or damaged elements. Their capacity to differentiate between normal and abnormal cells is a fundamental aspect of the immune system's protective mechanisms. (Source: Regenerative Medicine Partnership in Education, Https://www.sepa.duq.edu/regmed/immune/macrophages.html).

Lymphocytes in innate immunity include natural killer cells. These function by injecting cytotoxic chemicals into the invader, thereby killing it.

Third Line of Defense

Sometimes the intruder is new and has never been encountered. In this case, the immune system must create new antibodies to fight the intruder. This line of defense is known as adaptive immunity.

There is an interesting story told about the immune system. Onesimus was a slave to Cotton Mather, a Puritan minister in Boston during the 18th century. In 1712, a smallpox epidemic broke out in Boston. Onesimus told Cotton Mather about a practice that they had observed in Africa. They would take the pus of an infected person and, using a thorn, they would scratch it into the skin of an unaffected person. The introduction of smallpox gave the people immunity from the disease. Reactions varied from no reaction to a mild, non-fatal form of the disease. Onesimus saved numerous lives by telling his story and introduced the practice of inoculation into the United States. We mentioned how George Washington used the same practice on soldiers when smallpox broke out during the Revolutionary War.

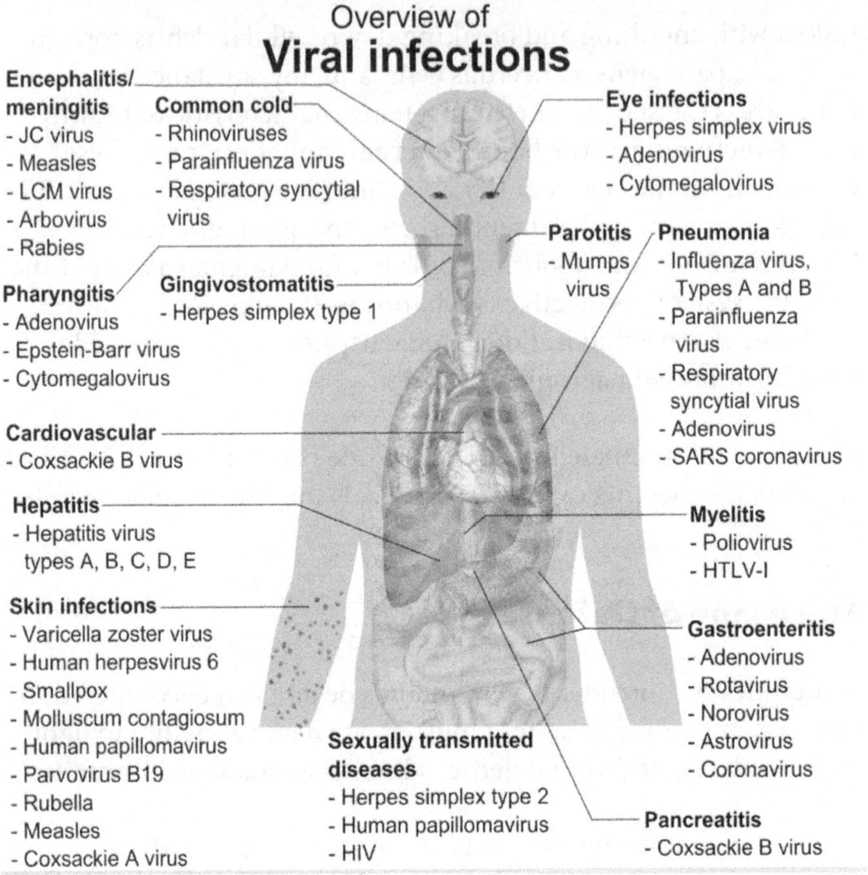

Public Domain

Viruses:

Some computer viruses are aptly named because they act like biological viruses. They attack the system and then try to replicate across an entire system or network. Early email viruses did just that. They read through the address book on the target system and then sent their code to everyone on the list. Some viruses are harmless irritations; others are highly destructive. To a large extent, systems have been developed to recognize an attack and eliminate it before it can do any harm. This is what the body does as well; if it recognizes an intruder, it will send out attackers who have been stored in memory for just such a case and destroy the invader. This is how inoculations work: by training the immune system to recognize a particular invader's code and then building antibodies to destroy it. These definitions are kept in the body's immune memory, ready to be released when the body senses a specific attacker. These antibodies can be manufactured very quickly. Some of these memory cells only last a short while; others will last a lifetime. This is why you need a "booster" shot for things like tetanus, although tetanus is a bacterial infection and not a virus.

Viruses do not have a cell membrane, cytoplasm, or chromosomes. They are made up of strings of DNA or RNA encased in a protein coat referred to as a capsid. Viruses cannot replicate themselves; they depend on the machinery of other cells to replicate. They enter the body through the nose, mouth, or breaks in the skin. Once in the body, they attack the cell, causing it to replicate the virus using the duplication machinery of the cell. Once the virus is duplicated, it is released from the cell and goes on to infect other cells.

The virus may have a match to the cell's receptors, like any molecule that the cell recognizes, not unlike a lock and a key. If that is the case, the cell readily welcomes the virus as acceptable and freely receives it. At this point, the deception is complete, and the viral information is sent to the replication machinery found in the cell, as if it were commanded to replicate a specific protein.

Once the viruses are fully developed, depending on their type, they either exit the cell by exocytosis (active transport from the cell) or lyse (break open) the host cell to spread themselves to neighboring cells and start the process over again.

There are viruses that may lie dormant for years before becoming active. HIV is one of those that may lie dormant for ten years before starting its replication. HIV contains both viral protein and human protein capsids (covers) called an envelope. The human protein helps disguise the virus from the immune system by deception. The viruses that have only a protein coat are called naked. The naked viruses cannot lie dormant because the immune system would destroy them before they could replicate.

Some viruses are called transforming viruses. In this case, the virus adds new information to the DNA. This type does not destroy the cell but transforms it. Warts are a type of transforming virus. This causes a wart to form on the skin. Once the wart grows to a certain size, it stops growing.

Interferons are a type of signaling protein that alerts nearby cells of an attack by a pathogen such as viruses, bacteria, parasites, or tumor cells. Interferons are part of a larger class known as cytokines. They are so named because of their ability to "interfere" with an attack. The interferons activate natural killer cells and macrophages.

> Interferon beta-1a and interferon beta-1b are used to treat and control multiple sclerosis, an autoimmune disorder. This treatment is effective for reducing attacks in relapsing-remitting multiple sclerosis and slowing disease progression and activity in secondary progressive multiple sclerosis - Paolicelli D, Direnzo V, Trojano M (14 September 2009). "R*eview of interferon beta-1b in the treatment of early and relapsing multiple sclerosis". Biologics: Targets & Therapy*

Bacteria

Unlike viruses, bacterial cells are alive. Because they are alive, they can be killed and cured. Viruses, however, can not be cured and can come back. Classified as prokaryotes, because they do not have a membrane-bound nucleus or any other membrane on its organelles.

There are good types and bad types of bacteria. The good bacteria is needed to break down food in the digestive track. Ask anyone who has taken antibiotics over an extended period and you will find that they commonly have digestive issues, because the antibiotics have killed the good bacteria. This can be cured by probiotics which contain live bacteria cultures available at most health food stores. Please check with a qualified health professional before starting any plan.

There is a lot of speculation from evolutionists that bacteria, that is resistant to antibiotics is proof of evolution. However, when examined it turns out not to be true. This is best explained by Kevin L. Anderson.

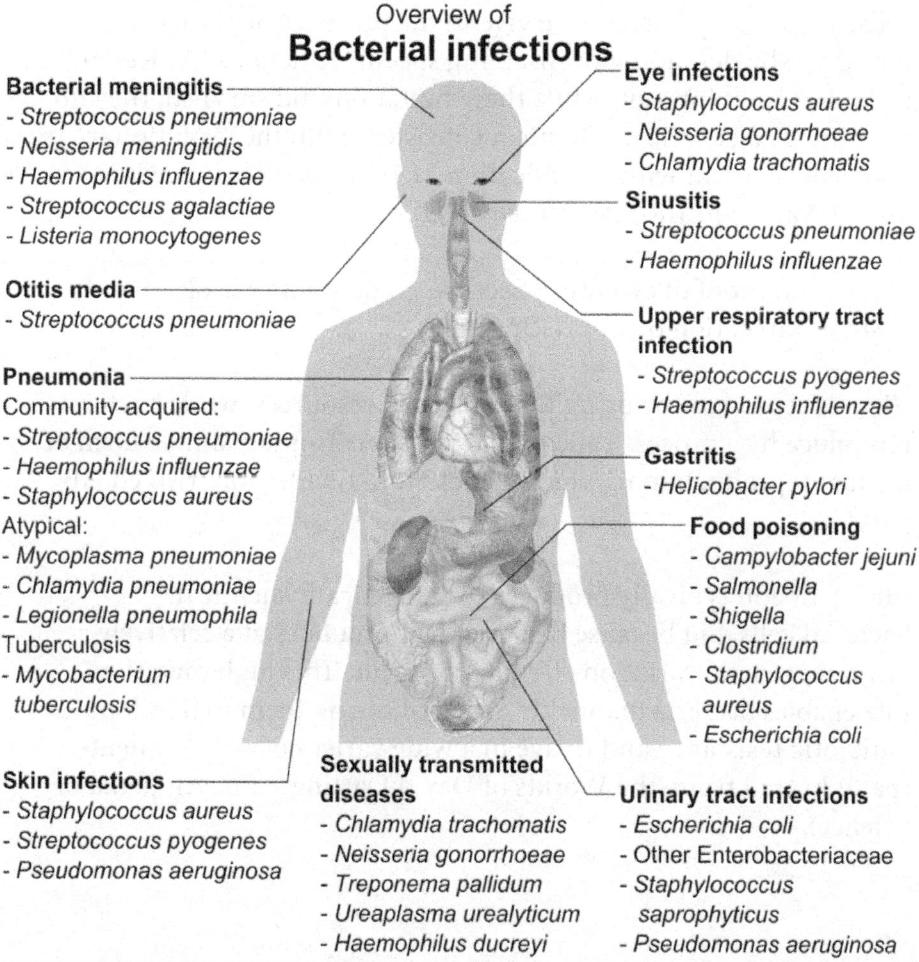

Many bacteria become resistant to antibiotics by acquiring genes through plasmids or transposons via a process known as horizontal gene transfer. However, while this process explains the spread of resistance genes among bacterial populations, it does not explain their origin. The initial emergence of antibiotic resistance is often attributed to mutations, but these mutations typically operate in ways that contradict evolutionary expectations. Rather than creating new functions or increasing complexity, such mutations usually reduce or eliminate the function of existing components—such as transport proteins, porins, enzyme activities, protein binding affinities, the proton motive force, or regulatory control systems. Though these mutations can be labeled "beneficial" because they help bacteria survive in the presence of antibiotics, they do so by degrading or disabling specific functions. As Kevin L. Anderson points out, while these mutations aid survival, they do not provide a genetic mechanism consistent with the evolutionary model of "descent with modification." (*Anderson, Kevin L. CRSQ Vol. 41 No. 4, pp. 318–326, March 2005*).

There is no proof of evolution because no new entity evolved; only a slight modification of an existing one.

Viruses reproduce by using the cells' own resources, while bacteria reproduce by a process called binary fission. This is a simple form of asexual reproduction in which a bacterium divides into two equal parts.

Binary fission typically produces two genetically identical bacterial cells, but because bacterial DNA mutates at a relatively fast rate, genetic variations frequently occur. This high mutation rate enables bacteria to quickly adapt, allowing them to develop antibiotic resistance and thrive in a wide variety of environments (paraphrased from The Worlds of David Darling – Encyclopedia of Science)."

Sternutation

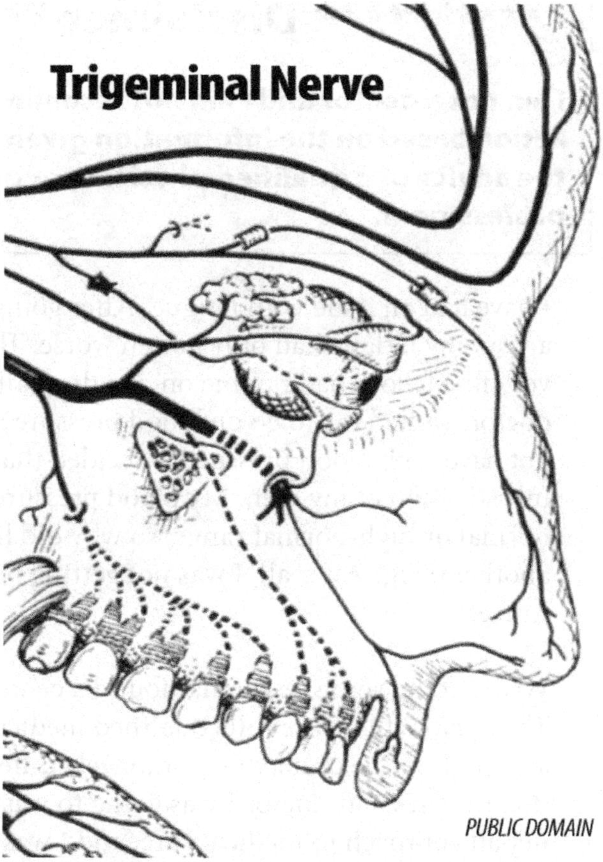

PUBLIC DOMAIN

One of the body's auto-responses to foreign objects is sternutation or sneezing. The purpose of a sneeze is to expel foreign objects from the nasal cavity. When enough stimulus is received from nasal hairs, it will call for the release of histamines. The histamines trigger the brain to induce a sneeze through the trigeminal nerve in the nasal cavity. Because of the force of a sneeze, a lot of the invaders can be expelled. It is said that when you sneeze the air coming out is traveling at about 100 miles per hour.

Why does pepper make you sneeze? The pepper contains an alkaloid of pyridine called piperine. Piperine is an irritant to the nasal mucosa which causing the message to be sent via the trigeminal nerve cause the brain to initiate a sneeze.

************* Disclaimer *******************

I am not a doctor and I am not recommending you take any action based on the information given in this book without the advice of a qualified physician or qualified health professional.

I have had an issue with fatigue. After going to a doctor for over a year, my fatigue had only gotten worse. The doctor, who is a very fine fellow, had put me on one drug after another. When the doctor wanted me to go on blood pressure medicine, and I did not have high blood pressure, I decided that I had better do a little investigation of my own. My blood pressure was always in the normal or high-normal range, so why did he want me to take yet another drug? After all, I was not getting better, but I was feeling worse.

While doing some research, I found a center specializing in fatigue. The doctors there are fully qualified medical doctors, but their approach to the problem is completely different. I had a feeling that the previous doctor I was going to was taking a one-size-fits-all approach to medical care, and I was getting worse, not better. The doctor at the Fatigue Center ordered 32 vials of blood for multiple tests. The results revealed a whole lot of things. I was low on testosterone, vitamin D, and many other things. In a very short time, following the doctor's advice, I began to feel much better. One thing the center doctor advised is not to take statin drugs (cholesterol-lowering drugs). After looking into it, I began to understand why.

Because doctors are taught an evolutionary perspective on the human body, they take an approach that the body is not capable of healing itself. Therefore, the approach is to look at the symptoms and, based on their medical training, prescribe a drug to deal with the issue. Unfortunately, if you look at the potential side effects of some synthetic drugs, it almost seems worse than the cure. The question not asked is: What caused this symptom to appear? This approach takes more research and time than doctors have to give to individual patients. I do not blame doctors; they are trained to diagnose illness. They are not trained to find and correct the cause of the problem. Drug companies have a lot invested in the training of doctors. I am not blaming drug companies. The real issue is money. Follow the money, and you will see what is behind the increase in drug advertising. I truly believe that doctors and the scientists involved in the use and creation of drugs are genuinely looking for solutions. Money is the underlying cause of this syndrome. Yes! The drug companies have a right to make a profit from their investment in research. The problem is that solutions that do not involve patented drugs are very inexpensive, and drug companies cannot make money on those. The research grants and money go to fund those research projects that will generate cash. That is a good business model if you want to stay in business. Research scientists, then, do not want to upset their benefactors by publishing work that may cause drug companies to lose money to inexpensive natural solutions. Please do not get me wrong. Drug companies have done a tremendous amount of good. This is a great model when dealing with inanimate objects like cars and computers. It is a very dangerous model when dealing with human life because low-cost solutions are abandoned in favor of drugs. Sadly, the same scenario is true in evolutionary "science" and "global warming" research. If you publish findings that go against the holder of the purse strings, you will lose your funding, your standing, and most likely your job. (See Expelled: No Intelligence Allowed.) This is no way to establish truth (or the answer), but maybe truth or an answer is not the goal...

I came across an article by Lisa Nelson RD, *Cholesterol Too Low. I asked her if I could include it in the book. She graciously allowed me to publish this article, you can find her at: Http://www.lisanelsonrd.com, please visit her site.*

> Cholesterol is often viewed as "bad" these days and many people are doing everything they can to lower

cholesterol levels as much as possible. This isn't necessarily a good thing.

Cholesterol is essential for many normal bodily functions. Enzymes use cholesterol to produce vitamin D, steroid hormones (estrogen, progesterone, testosterone), stress hormones, and bile acids for digestion. Cholesterol forms a membrane that surrounds all cells and is also a critical part of regenerating damaged endothelial cells (inner layer of blood vessel walls).

For most individuals, if you do not eat enough dietary cholesterol the liver produces the cholesterol needed for bodily functions. The amount of cholesterol in your diet determines how much the liver produces.

Cholesterol is essential for optimal health and pushing levels too low can cause problems. Back in 1994, the American Heart Association issued a statement noting an increase in deaths from trauma, cancer, hemorrhagic stroke, respiratory diseases, and infectious diseases in individuals with cholesterol levels less than 160 mg/dl. It's worth noting that a large number of these deaths seemed to be due to poor health unrelated to the low cholesterol levels.

However, since then many studies are linking low cholesterol levels with depression, suicide, impulsivity, aggression, and anxiety when levels drop below 160 mg/dl. One psychologist suggested having too low cholesterol alters brain cell function and the brain cells with low cholesterol levels may have fewer receptors for the mood-elevating neurotransmitter serotonin. This could be the reason low cholesterol levels increase rates of depression.

With the rapidly rising use of medications, such

> as statin drugs, I'm noticing individuals reporting cholesterol levels that are dropping into potentially dangerous territory. Work with your physician to ensure you maintain healthy cholesterol levels appropriate for you.

I am thankful that people like Lisa Nelson recognize the issues with statin drugs. Lisa's job is to recommend dietary suggestions to help the body heal itself. Thank you, Lisa.

Let's consider cholesterol and how it relates to the immune system and white blood cells known as macrophages. In Lisa's article, she mentions that cholesterol has many functions, like helping enzymes in generating brain cells, Vitamin D, promoting steroid hormones, and much more. Further, she notes that lowering cholesterol too low can be dangerous, causing maladies like depression, cancer, and other serious problems. Since cholesterol has so many benefits, why all the talk about "bad" LDL cholesterol?

LDL cholesterol is related to the problem of atherosclerosis or hardening of the arteries. The problem exists because we live in a world filled with chemicals, radiation from the sun, insufficient vitamins and minerals in our food, improper diet, lack of exercise, and so on. This causes a rise in the occurrence of what is commonly called "free radicals" in our body.

Free radicals "oxidize" molecules in the body, similar to what happens to a car's finish if left in the sun without protection; the finish gets "oxidized." If an LDL cholesterol molecule gets oxidized, its chemical signature will change. The immune system recognizes oxidized LDL cholesterol as a foreign substance in the bloodstream, and the immune system is activated to eliminate the intruder. The white blood cell macrophage sees the oxidized LDL and pulls it out of the bloodstream and into the wall of the artery. This is the beginning stage of what is known as a "foam cell" because it looks similar to foam. After a while, the foam cell begins to secrete cytokines (proteins that are important in cell signaling). The cytokines alert the immune system to activate. The immune system wants to section off these invaders, and so it lays down connective tissue similar to bone. This is the beginning of the dreaded "hardening of the arteries."

Doctors are taught to use statin drugs to lower the troublesome LDL cholesterol. However, this does not address the reason that LDL cholesterol is clogging the artery in the first place. Lowering the LDL levels only masks the greater problem of free radical damage. The answer is a diet of antioxidant-rich foods, vitamins, minerals, and sufficient levels of exercise, among other things. However, you must get direction from a fully qualified physician before you attempt to implement such a plan.

Just like the eye and many other critical functions in the body, the immune system depends on several systems to operate: thymus, spleen, lymph system, bone marrow, white blood cells, antibodies, hormones, just to name a few. Our description of the immune system only introduces you to the many things that go on in the immunity function of our bodies. Dr. Geoffrey Simmons said it best; it is a "*whole package phenomenon.*" The evolutionist has a difficult task ahead: explaining how this marvel of intricate and synergistic design came about by undirected chance mutations. For the immune system to work properly, all of the pieces had to be there from the beginning, or no human would survive the relentless attacks of foreign substances entering our body every minute of the day. Without a complete immune system, you would be sending an army to stop invaders, but they don't know who they are, what they look like, or how to defeat them. They would have no direction and no weapons to work with and would soon be overtaken by the enemy. No! Our bodies would not have survived without the marvel of the immune system.

Inclusion Body Myositis (IBM)

I include this issue because I am acquainted with someone who has died from this disease.

This type of disease is categorized as inflammatory myopathies. There are two types of this issue: one comes on without warning (sIBM), and the other is genetic: hereditary inclusion body myopathy (hIBM). It is characterized by slow wasting and weakness of the arms and legs. The malady is not fatal but can be potentially fatal if the disease progresses and a condition known as dysphagia appears. Dysphagia is a medical term for difficulty in swallowing. This can lead to very serious problems and even death; however, this is not the case with every sIBM patient.

sIBM symptoms usually increase with age, typically after the age of 50, but can occur at ages 35 to 90. The causes of sIBM are currently unknown, which makes it quite difficult to diagnose properly. There is a lot of confusion when attempting to diagnose sIBM because not all patients have a similar biopsy."

> In sporadic inclusion body myositis [MY-oh-sigh-tis], two processes, one autoimmune and the other degenerative, appear to occur in the muscle cells in parallel. The inflammation aspect is characterized by the cloning of T cells that appear to be driven by specific antigens to invade muscle fibers. The degeneration aspect is characterized by the appearance of holes in the muscle cell vacuoles, deposits of abnormal proteins within the cells and in filamentous inclusions (hence the name inclusion body myositis).
> - Wikipedia contributors. "Inclusion body myositis."

It has been determined that an autoimmune response (where the immune system attacks the body) as a cause is not tenable. It is rejected because of the many tests that have been done using immunotherapy. These treatments have failed to produce the desired results.

There have been many theories that have tried to explain what causes sIBM, but nothing has been sufficiently proven so far. Part of the problem is the fact that sIBM is often misdiagnosed as polymyositis (PM) or dermatomyositis (DM).

Polymyositis is the chronic inflammation of the muscles. An unknown autoantigen causes the disease of muscle tissue, which is affected by cytotoxic T cells (more on T cells in chapter 6). This disease is normally treated with corticosteroids and specialized exercise.

> Dermatomyositis is characterized by inflammation of the muscles and the skin. While DM most frequently affects the skin and muscles, it is a systemic [all parts of the body] disorder that may also affect the joints, the esophagus, the lungs, and the heart.
>
> ...The pain may resemble the type experienced after strenuous exercise. Some dermatomyositis patients have little pain, while in others (esp. in juvenile dermatomyositis), the pain may be severe. It is important to remember that this condition varies from person to person in many ways. Also in many cases muscle may deteriorate and render the patient temporarily paralyzed unable to walk, run, get out of bed, or even swallow food and liquids.
> - Wikipedia contributors.
> Dermatomyositis." Wikipedia.

Just like polymyositis, dermatomyositis is characterized by muscle deterioration; sIBM makes it difficult to distinguish between the other two, and therefore it is hard to properly diagnose. The difference is that sIBM does not respond to therapy. Typically, the disease is asymmetrical, so weakness may not be the same on both sides of the body.

One theory is that the muscle fibers are surrounded by T cells and other immune system cells but are not invasive. However, the chemicals secreted by the immune cells cause injury to the muscle cells.

In polymyositis and sporadic inclusion body myositis (sIBM), the commonly accepted model of immune-mediated myofiber injury involves cytotoxic T cells directly invading muscle fibers. However, this represents only a small portion of the overall disease process. In reality, a much greater number of myofibers are surrounded and often displaced by T cells and other immune cells without being directly invaded. This observation has led to speculation that the injury to myofibers may result not only from direct invasion but also from molecules secreted by the numerous non-invading immune cells. (Inflammatory Myopathies: Disease Mechanisms – Steven A. Greenberg)

A notable trait of muscle impacted by inclusion-body myositis (IBM) is the irregular accumulation, clumping, and misfolding of various proteins. This pathological aspect bears a striking resemblance to the protein irregularities found in the brains of individuals with Alzheimer's disease. This indicates that, despite the differing origins of IBM and neurodegenerative conditions such as Alzheimer's, there may be comparable underlying mechanisms at play. (Askanas V, Engel – *Inclusion-Body Myositis and Myopathies: Different Etiologies, Possibly Similar Pathogenic Mechanisms, Current Opinion in Neurology*, 2002)

Statistically, patients with PM and DM have these results: 45-55% do fine and eventually get off drugs, 35% have mixed results, and about 10% do not fare well. IBM patients do not respond to drug therapy, and the disease is slowly progressive. There is currently no cure for the disease.

You have a choice: believe that this intricate system of building the cells, communicating, and organizing an attack on the immune system is the work of undirected chance happenings, as the evolutionists want you to believe, or that it was designed from the beginning.

POLITICS IN
HEALTH CARE

Politics in Health Care (excerpt from Book 1)

If you think that the politicians, drug companies, and food companies have your best health interests in mind, I have a rude awakening for you: they do not! Politicians need money to get re-elected; food and drug companies need customers for their products, and universities need money for their existence. Do you begin to see the potential for bad policies and biased research? Drug companies are very willing to give massive amounts of money to politicians and universities that will follow their advice. Doctors are given virtually no training on nutrition (I have heard MDs say they had less than one day of nutritional training in medical school). How can a doctor give advice on nutrition? What they are taught in medical school is how to address an illness with drugs, not how to reverse a disease. It is my experience that doctors do not concern themselves with why a patient has a particular disease, only with 'how can I relieve the symptoms?

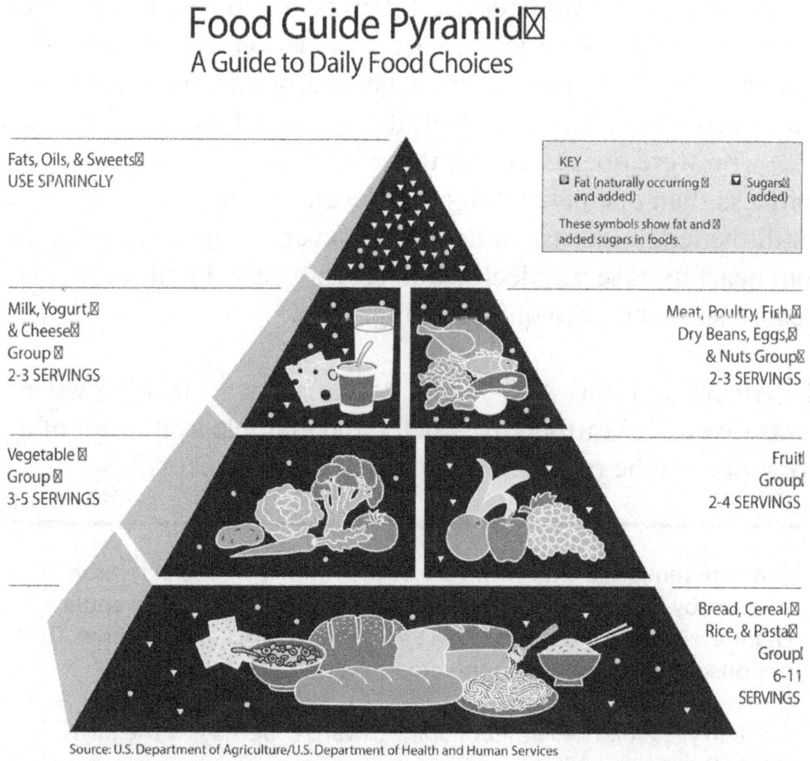

Source: U.S. Department of Agriculture/U.S. Department of Health and Human Services

Do you think studies that come out of universities or researchers that receive large amounts of funding from food and drug companies will be unbiased? The obvious answer is a resounding, **NO!**

> For every complex problem there is an answer that is clear, simple, and wrong. - H. L. Mencken

Nina Teicholz, in her book *The Big Fat Surprise: Why Butter, Meat, and Cheese Belong in a Healthy Diet*, points out that the foods we have been told are bad for us are the very foods that can reverse obesity, diabetes, and heart disease. I can personally attest to that fact. There is a growing number of people who are experiencing the benefits of returning to a diet our forefathers ate before all the misinformation about cholesterol and saturated fats. Teicholz goes on to say that in the early 1900s, there were only a few who died of heart disease, but by the 1950s, heart disease became the number-one killer. In her book, she cites that in 1961, when the American Heart Association (AHA from now on) recommended a low-fat, low-cholesterol diet, only one in seven Americans were obese. By 2001, that number had leaped to one in three who were obese. During the same time, diabetes had risen from less than one percent to over eleven percent, and heart disease is still the leading cause of death. However, the number of deaths from heart disease has declined in recent years; I believe this is because so many have quit smoking.

The American Heart Association got its real start in 1948 when it was given $1.5 million by Procter and Gamble as a result of a promotion on the radio show Truth or Consequences.

> The AHA remained small until the 1940s when it was selected for support by Proctor & Gamble, via their PR firm, from a list of applicant charities. Proctor & Gamble gave $1.5 million from its radio show, Truth or Consequences, allowing the organization to go national.
>
> Tye, Larry (1998). The Father of Spin: Edward L. Bernays & the Birth of Public Relations. p. 74.

Not surprising that the AHA said that Crisco (Crystallized Cottonseed Oil, made by Proctor & Gamble) is healthier than butter. We now know that trans-fat from vegetable oils is very unhealthy even deadly!

"Follow the money".

Ancel Keys, a name that will live in infamy, will become known as one of the most villainous health researchers in American history. Once everyone is made aware of Keys' biased and deceptive practice of leaving out major parts of his research to prove his very flawed hypothesis that dietary saturated fat causes cardiovascular heart disease and should be avoided. Over the years, anyone who dares to contradict this premise has been silenced by being cut off from grants and having their influence disappear.

> "So what is the answer to increased health and vigor? Eating more or fewer foods is not the answer. The key lies in knowing what to eat. You need to know how to obtain the essential micro and macronutrients and the amount needed daily. This is a very important point that the food pyramid system fails to consider. The reason for this is that current nutritional thinking pushes a philosophy of food groups and not the essential micro and macronutrients you need. When you are eating to get sufficient nutrients, you will not ask how many food servings to eat but the amount of the actual micro and macronutrients needed."
>
> Luan Pho. Health and Vitality Truths . Luan Q. Pho, M.D., P.A.

Death by Food Pyramid: How Shoddy Science, Sketchy Politics and Shady Special Interests Ruined Your Health **by Denise Minger** is another book which is a good read. Here is a brief description of the book on Amazon:

"Warning: Shock and outrage will grip you as you dive into this one-of-a-kind exposé. Shoddy science, sketchy politics, and shady special interests have shaped American Dietary recommendations--and destroyed our nation's health--over recent decades. The phrase "death by food pyramid" isn't shock-value sensationalism, but the tragic consequence of following federal advice and corporate manipulation in pursuit of health.

In Death by Food Pyramid, Denise Minger exposes the forces that overrode common sense and solid science to launch a pyramid phenomenon that bled far beyond US borders to taint the eating habits of the entire developed world"

Minger points out that the "experts," the "highly educated," and the "conventional wisdom" that promoted the idea that saturated fats will clog your arteries and other misinformation are completely misguided. The consequences of following the "Food Pyramid" are dire at best.

"… these guidelines might actually have had a negative impact on health, including our current obesity epidemic. [There's a] possibility that these dietary guidelines might actually be endangering health is at the core of our concern about the way guidelines are currently developed and issued."

– Paul Marantz, Albert Einstein College of Medicine, *American Journal of Preventative Medicine*

Minger followed a vegetarian diet of 80/10/10 with alarming consequences. She then goes on to tell the horrific story of the politicization of the food pyramid.

The Swedes were in need of cheaper food. Their focus was on cost rather than disease prevention. In an attempt to help alleviate the high cost of food in Sweden, in 1974, Anna-Britt Agnsäter put together a food pyramid to visualize how to eat well while cutting costs. At the base were dairy, grains, pasta, and potatoes; in the middle were vegetables and fruits; at the top were meat, fish, and eggs. Twenty years later, the USDA came out with its infamous food pyramid, which in many ways mirrored Agnsäter's pyramid.

You will find that some of the food groups lead to a multitude of health issues. This has been proven by statistics that show an enormous increase in obesity, diabetes, and even new maladies of the brain that are on the rise. Based on the results, following the recommendations given by the establishment would be insane. You know what they say about insanity: "Insanity is doing the same thing over and over while expecting different results.

> "I diagnose by a clinical picture. I believe in clinical experience and clinical picture, much more than I believe in science. Our science has lost its way. ***More than 90% of scientific studies are commercially funded. They are deceptive.*** And the majority of us have no tools, and no training, to really decipher whether the study was designed properly, conducted properly, and then analyzed properly. It's very difficult. So there's a huge amount of information that leads you away from the truth. So I work with real people on a daily basis. I see what works and what doesn't work in real life. And if I need some scientific backing for any of those truths that I've discovered, I will find it. And usually those scientific studies are kosher... are correct. I believe in Mother Nature. I believe in real life, and real people." [**my emphasis**]
>
> - Dr. Natasha Campbell-McBride
> - *Gut-Brain Secrets, Good Food, Bad Food*

Following list is from a talk by Dr. Michael Eades author of *Protein Power.*
Https://www.youtube.com/watch?v=wL5513xKw1k

- **1978** High Fructose Corn Syrup enters the market
- **1980** Obesity stable at 12-14% since 1960
  USDA releases low-fat guidelines
- **1984** NIH (National Institutes of Health)
  Lipid Research Clinic Study
  Anthony Gotto - President of AHA (American Heart Association)
  said "We will conquer atherosclerosis by the year 2000"

- **1986** NIH & AHA establish the NCEP (National Cholesterol Education Program)
  FDA says "no conclusive evidence" that sugar causes chronic disease
- **1987** Mevacor - first statin approved
- **1988** Surgeon General "highest priority given to reducing fat in take" C. Everett koop

Since the late 1970s, we have seen a huge increase in the number of diseases and those who have them. Obesity, diabetes, heart disease, stroke, and new diseases of the brain are a pandemic of biblical proportions. I believe this is a result of the recommendations of various medical associations and bad science.

I went to the National Institutes of Health's website and copied what they said in 1988... I disagree vehemently!

> High intake of total dietary fat is associated with increased risk for obesity, some types of cancer, and possibly gallbladder disease. Epidemiologic. clinical, and animal studies provide strong and consistent evidence for therelationship between saturated fat intake, high blood cholesterol, and increased risk for coronary heart-disease. Conversely, reducing blood cholesterol levels reduces the risk for death from coronary heart disease. Excessive saturated fat consumption is the major dietary contributor to total blood cholesterol levels. Dietary cholesterol raises blood cholesterol levels, but the effect is less pronounced than that of saturated fat. While polyunsaturated fatty acid consumption, and probably monounsaturated fatty acid consumption, lowers total blood cholesterol, the precise effects of specific fatty acids are not well defined. - NIH.gov - The Surgeon General's report on Nutrition and Health: Summary and Recommendations (1988)

If there was any evidence that the dietary guidelines were detrimental to our health, then the CDC's stats would prove it was bad advice. The guidelines were introduced in 1980, at which point we saw a dramatic increase in weight and obesity. Did we suddenly start eating more fat and cholesterol? Was it butter, red meat, and dairy? NO! The recommendation was to eat low fat, add grains, fruits, and vegetable oils, and eat less meat, eggs, and dairy, which, for the most part, we did. Why did our ancestors before the 1900s not have an epidemic of heart disease, diabetes, and other metabolic issues? Our ancestors ate meat, eggs, and dairy.

"This [pharmaceutical] industry uses its wealth and power to co-opt every institution that might stand in its way, including the US Congress, the Food and Drug Administration, academic medical centers and the medical profession itself " - Dr. Marcia Angell, former editor of the New England Journal of Medicine - *The Truth about the Drug Companies: How They Deceive Us and What to Do about It.*

The USDA (United States Department of Agriculture) during this time (1980s) wanted to increase grain consumption. This goal could easily be reached if everyone followed the advice to eat a diet low in fat and high in carbohydrates. Since that time, we have seen an explosion of obesity and related health problems.

"The main reason we take so many drugs is that drug companies don't sell drugs, they sell lies about drugs. This is what makes drugs so different from anything else in life...Virtually everything we know about drugs is what the companies have chosen to tell us and our doctors... the reason patients trust their medicine is that they extrapolate the trust they have in their doctors into the medicines they prescribe. The patients don't realize that, although their doctors may know a lot about diseases and human physiology and psychology, they know very, very little about drugs that hasn't been carefully concocted and dressed up by the drug industry.."

Deadly Medicines and Organized Crime - Peter Gotzsche

Statin Adverse Effects: A Review of the Literature and Evidence for a Mitochondrial Mechanism - Beatrice A. Golomb, M.D., Ph.D. and Marcella A. Evans - Https://www.ncbi.nlm.nih.gov/pmc/articles/PMC2849981/

The following is a digest of: Statin Adverse Effects (AEs) Supported by Evidence from Randomized Controlled Trials (RCTs)

- Muscle - increase in myositis (inflammation of muscle)
- Cognition - Muscle and brain are classically affected in coenzyme-Q10-deficiency mitochondrial syndromes.
- Cancer - The sole randomized trial in the elderly (age >70) showed significant increase in incident cancer with statin use relative to placebo (HR 1.25, 95% CI 1.04-1.51, p=0.02)
- Liver - RCT data of LFT (liver function test) elevations are buttressed by many cases of statin hepatopathy (Disorder of the liver)arising alone or in concert with statin rhabdomyolysis (potentially fatal disease that destroys skeletal muscle)
- Hemorrhagic Stroke - This extends evidence from observational data showing higher hemorrhagic stroke risk with low cholesterol
- Blood Glucose - High-dose statins led to statistically significant increase in glycemia in the PROVE-IT–TIMI trial
- Sleep - Also, there are case reports and case series of sleep problems and nightmares; and reports of sleep AEs in clinical trials of statins

Cholesterol and statins: Sham science and bad medicine

Michel de Lorgeril, M.D., is a cardiologist, nutritionist and researcher at France's National Center for Scientific Research (CNRS) and a member of the European Society of Cardiology

A description from his book on Amazon.com:

He speaks out against this collective regulatory insanity, based on biased, truncated and often falsified studies.
- The theory that cholesterol 'blocks arteries', that it causes heart attacks and strokes is an illusion that does not stand up to any physiological, experimental, epidemiological or clinical argument.

- Contrary to current dogma, statins do not reduce mortality, and this book provides irrefutable proof of this.

- These drugs, which can lead to diabetes and cancers, can also cause severe muscle damage, cognitive impairment and sexual dysfunction.

Doctors and patients need good quality scientific evidence in order to make informed decisions. But for decades now they have been misled by directives based on corrupt clinical studies, which have been knowingly altered to exaggerate the benefits of their molecules and to minimise their secondary effects. When their findings are carefully analyzed, nothing remains of their claims.

By far the most serious side effect, says Michel de Lorgeril, is that these prescriptions convey a false sense of security and prevent the population from adopting healthier lifestyle measures. Measures whose efficacy in the prevention of heart attacks and cardiovascular disease has been proven - and which are presented in this book

The Truth About the Drug Companies: How They Deceive Us and What to Do About It - Dr. Angell

(From a description of the book on Amazon)

During her two decades at "The "New England Journal of Medicine, Dr. Marcia Angell had a front-row seat on the appalling spectacle of the pharmaceutical industry. She watched drug companies stray from their original mission of discovering and manufacturing useful drugs and instead become vast marketing machines with unprecedented control over their own fortunes. She saw them gain nearly limitless influence over medical research, education, and how doctors do their jobs. She sympathized as the American public, particularly the elderly, struggled and increasingly failed to meet spiraling prescription drug prices. Now, in this bold, hard-hitting new book, Dr. Angell exposes the shocking truth of what the pharmaceutical industry has become-and argues for essential, long-overdue change.

Currently Americans spend a staggering $200 billion each year on prescription drugs. As Dr. Angell powerfully demonstrates, claims that high drug prices are necessary to fund research and development are unfounded: The truth is that drug companies funnel the bulk of their resources into the marketing of products of dubious benefit. Meanwhile, as profits soar, the companies brazenly use their wealth and power to push their agenda through Congress, the FDA, and academic medical centers.

Zeroing in on hugely successful drugs like AZT (the first drug to treat HIV/AIDS), Taxol (the best-selling cancer drug in history), and the blockbuster allergy drug Claritin, Dr. Angell demonstrates exactly how new products are brought to market. Drug companies, she shows, routinely rely on publicly funded institutions for their basic research; they rig clinical trials to make their products look better than they are; and they use their legions of lawyers to stretch out government-granted exclusive marketing rights for years. They also flood the market with copycat drugs that cost a lot more than the drugs they mimic but are no more effective.

The American pharmaceutical industry needs to be saved, mainly from itself, and Dr. Angell proposes a program of vital reforms, which includes restoring impartiality to clinical research and severing the ties between drug companies and medical education. Written with fierce passion and substantiated with in-depth research, "The Truth About the Drug Companies is a searing indictment of an industry that has spun out of control.

see also the video by Jack Bresler: *Swimming in Pills*:
Https://vimeo.com/31058254

We are living in an age of deceit. It is getting more difficult to know whom to trust. Many ads promise miraculous results, but are most times outright fraud. If a product has a 30-day guarantee, and you get your money back, however, they have had the use of your money for 30 days (a small interest-free loan).

> "It is simply no longer possible to believe much of the clinical research that is published, or to rely on the judgment of trusted physicians or authoritative medical guidelines. I take no pleasure in this conclusion, which I reached slowly and reluctantly over my two decades as an editor of The New England Journal of Medicine." - Harvard Medical School's Dr. Marcia Angell

Drug companies distort the facts by trickery, promoting that their product will reduce the risk of death by some large percentage when, in fact, it is less than 2%. David Diamond points out that an ad for Lipitor® states that it will reduce the risk of fatal and non-fatal heart attacks by 36%. This is accomplished by dividing the factual risk when taking a placebo by the actual risk when taking the drug. In this case, 98.1% did not die of a heart attack when taking the drug, and 97% did not die when taking the placebo. The difference is 1.1%. Take the 1.1% difference and divide it by the 3%, and you get 36%. In reality, the only benefit of taking the statin over the placebo is minuscule, yet the ad implies a much greater benefit.

Food companies proclaim that their food is healthy, and some even have the heart-healthy label, even though it is full of empty carbohydrates and sugar. We have illustrated what happens when sugars and carbohydrates are processed by the body. Food companies try to say that all calories are the same; 100 calories in a can of soda is equivalent to 100 calories in a vegetable. This is provably false! Your body will disagree with that proposal. The body does not treat every calorie the same, as we will illustrate later!

AHA's (American Heart Association) income, which amounts

to millions of dollars a year, comes from its "Heart-Check Certification Program." For a fee, food companies can pay to use the "Heart Healthy" symbol on their products, a red heart with a white check mark. The designation can be found on products that are NOT heart healthy, based on what we now know about sugar and carbohydrates.

Trust Us We're Experts: How Industry Manipulates Science and Gambles with Your Future - by John Stauber, Sheldon Rampton

"You think that nonprofit organizations just give away their stamps of approval on products? Bristol-Myers Squibb paid $600,000 to the American Heart Association for the right to display AHA's name and logo in ads for its cholesterol-lowering drug Pravachol. SmithKline Beecham paid the American Cancer Society $1 million for the right to use its logo in ads for Beecham's Nicoderm CQ and Nicorette anti-smoking ads.

You think that if a scientist says so, it must be true? In the early 1990s, tobacco companies secretly paid thirteen scientists a total of $156,000 to write a few letters to influential medical journals. One biostatistician received $10,000 for writing a single, eight-paragraph letter that was published in the Journal of the American Medical Association. A cancer researcher received $20,137 for writing four letters and an opinion piece to the Lancet, the Journal of the National Cancer Institute, and The Wall Street Journal" (Quote from Amazon description of the book)

What about doctor error? I generally do not blame doctors; after all, they are "practicing medicine." The human body is an extremely complex machine, and knowing exactly how to deal with each instance is very difficult. Each one of us is distinct—our lifestyles, our backgrounds, and our genetics. Sometimes, doctors get it wrong due to outdated information from schooling; other times, it is due to faulty drug studies, and often it is because of incorrect guessing.

Only medical doctors can prescribe drugs. Pharmaceutical companies spend enormous amounts of money promoting their products to doctors. However, have you noticed how many drugs are advertised in the media today? You are bombarded by ads touting the latest miracle drug, and you are innocently told to "ask your doctor" about drug X. There is a method to the madness. If enough people talk to their doctors about drug X, it is a silent promotion of their product to doctors. "Word of mouth works!"

See the book The Big Fix: How The Pharmaceutical Industry Rips Off American Consumers (Public-affairs Reports) by Katharine Greider.

At the time I am writing this, you can read a report by Dr. Gary Null, M.D., which documents the issues and costs of misdiagnoses in terms of money and lives: www.newmediaexplorer.org/sepp/ Death by Medicine Nov 27.doc.

"For a modern disease to be related to an old fashioned food (e.g. red meat) is one of the most ludicrous things I have ever heard in my life" - Peter Cleave, Surgeon Captain (1906-1983)

"We have a multi-billion dollar industry that is killing people, right and left, just for financial gain. Their idea of research is to see whether two doses of this poison is better than three doses of that poison." — Dr. Glenn A. Warner, M.D. (1919-2000).

"The combined profits for the ten drug companies in the Fortune 500 ($35.9 billion) were more than the profits for all the other 490 businesses put together ($33.7 billion) [in 2002]. Over the past two decades the pharmaceutical industry has moved very far from its original high purpose of discovering and producing useful new drugs. Now primarily a marketing machine to sell drugs of dubious benefit, this industry uses its wealth and power to co-opt every institution that might stand in its way, including the US Congress, the FDA, academic medical centers, and the medical profession itself." - Dr. Marcia Angell - *The Truth About the Drug Companies* - Dr. Angell is a former editor in chief of the prestigious New England Journal of Medicine.

From the **bmj** (the British Medical Journal:)

Big pharma often commits corporate crime, and this must be stopped BMJ 2012; 345 doi: Https://doi.org/10.1136/bmj.e8462 (Published 14 December 2012) BMJ 2012;345:e8462

Peter C Gøtzsche, professor, Nordic Cochrane Centre, Rigshospitalet, Blegdamsvej

When a drug company commits a serious crime, the standard response from the industry is that there are bad apples in any enterprise. Sure, but the interesting question is whether drug companies routinely break the law.

I googled the names of the 10 largest drug companies in combination with the term "fraud" and looked for offenses on the first page for each company. The most common recent crimes were illegal marketing by recommending drugs for non-approved (off label) uses, misrepresentation of research results, hiding data on harms, and Medicaid and Medicare fraud.1 All cases were related to the United States and involved huge settlements or fines, exceeding $1bn (£620.6m; €769m) each for four companies.

It was easy to find additional crimes committed by these same companies and committed outside the US. As the crimes were widespread and repetitive, they are probably committed deliberately—because crime pays. Pfizer, for example, agreed in 2009 to pay $430m to resolve charges related to illegal marketing of gabapentin (Neurontin), but as sales were $2.7bn in 2003 alone, and as about 90% was for off label use...

Dr. Robert Levy, National Heart Lung and Blood Institute director stated during the hearings:

> "With cholesterol the issue is a little more murky. We have no doubt from the vast amount of epidemiological data available that elevated [blood] cholesterol is associated with an increased risk of heart attack, especially some specific types of … cholesterol. We have no doubt that cholesterol can be lowered by diet, and/or medication in most patients. **Where the doubt exists is the question of whether lowering cholesterol will result in a reduced incidence of heart attack; that is still presumptive**. It is unproven [in clinical trials], but **there is a tremendous amount of circumstantial evidence**… .

There is no doubt that cholesterol can be lowered by diet in free living populations. It can be lowered by 10 to 15 percent. **The problem with all these trials is that none of them have showed a difference in heart attack or death rate in the treated group.**" (my emphasis)

It is clear that Dr. Levy had serious doubts about the question of the lipid hypothesis and disease. In the final analysis, the committee bent to special interest groups and political contentions. This is often the cause of many government programs gone astray, ending in actually harming the public it is tasked with helping. During this time (1977), there was a rush to complete the task of the McGovern committee, and McGovern said:

"Senators do not have the luxury that the research scientist does of waiting until every last shred of evidence is in,"

This rush to judgment has caused an immense amount of pain and suffering to the majority of those who followed their ill-advised nutritional advice, as evidenced by the increase in ill health and disease. You can read the details of the of this committee at: Https://www.ncbi.nlm.nih.gov/pmc/articles/PMC3910043/
I refer you to the article previously quoted from Brown University:

Article · May 2016 Richard Lear - Brown University

The Root Cause in the dramatic rise of Chronic Disease

See discussions, stats, and author profiles for this publication at:
Https://www.researchgate.net/publication/303673576

What has happened since we followed the advice from the McGovern commission in the late 1970s?

Autism (2,094%)
COPD (148%)
Sleep apnea (430%)
ADHD (819%)
Depression (280%)
Osteoarthritis (449%)
Inflammatory bowel (120%)
fibromyalgia (7,727%)
hypothyroidism (702%)
Alzheimer's (299%)
diabetes (305%)
celiac disease (1,111%)
asthma (142%)
youth bipolar disease (10,833%)
upus (787%)
chronic fatigue (11,027%)
multiple sclerosis (117%)

According to the article, Richard Lear attributes the increase in disease to peroxynitrite. Peroxynitrite is formed when l-arginine is not converted to nitric oxide (NO). This conversion causes inflammation, setting off a cascade of events that leads to the death of the mitochondria. When superoxide and nitric oxide radicals exist simultaneously in the cell, trouble begins; necrosis or apoptosis of the cell will be the outcome. All of the above diseases have been associated with peroxynitrite.

The mitochondria are responsible for many functions: energy production, calcium homeostasis, and the creation of new complex molecules out of simpler, smaller precursors (known as the biosynthesis pathway). Mitochondrial dysfunction is associated with many diseases: diabetes, stroke, atherosclerosis, and many others. It is my hope that you see that health is a very complex subject, one we could not begin to address in any great detail. It is my contention that drugs are seldom the answer. We must first look at the foods we consume and the toxic environments we find ourselves in, both physically and mentally.

If you want a detailed scientific explanation see: Https://www.sciencedirect.com/topics/pharmacology-toxicology-and-pharmaceutical-science/peroxynitrite

A

Comparison Chart

- Creation
- Intelligent Design
- Theistic Evolution
- Progressive Evolution
- Darwinian Evolution

Used with permission from

by J.D. Mitchell, P.E.

www.creationengineeringconcepts.org

	Young Earth CREATIONISM
a. Who are the proponents?	Creation Scientists
b. What is the history?	Since Creation ~6000 yrs
c. What is the public acceptance?	Confusion
d. Where did life come from?	Created By God
e. How about life form variety?	Information within Kinds
f. How did man originate?	From Dust at Creation
g. How many races are there?	One - Human Race
h. How about origin of the earth & universe?	Created in 6 days
i. How are days of creation explained?	Ordinary 24 hour days
j. What is the age of the earth?	~ 6000 years
k. What is the age of the universe?	~6000 years
l. How about dinosaurs?	Day 6 with Man
m. How about death, disease & suffering?	Punishment for Sin
n. Is there a moral authority?	The Bible
o. What is the view of the Bible?	Inerrant & Authoritative
p. What about Noah's Flood?	Cataclysmic & Worldwide
q. How are the geologic layers explained?	Noah's flood
r. What does the fossil record tell us?	Judgment & Flood
s. How about Noah's ark?	Large barge-like vessel

Young/Old	Old Earth	Old Earth	Old
Intelligent Design	**Theistic Evolution**	**Progressive Creation**	**Darwinian Evolution**
I. D. Movement	Churches & Theologians	Day Age Scientists and theologians	Secular World
Natural Theology of William Paley (1802)	Since Augustine (430 AD)	Since ~ 1800	Since Aristotle
Confusion	Confusion	Confusion	Confusion < 20%
From a designer	God created it then it evolved -	God created life in spurts	Accidentally from Chemicals
NCA	natural selection and mutations	God created separately	Natural Selection & mutations
NCA	From ape-like ancestors	God Made man supernaturally	From ape-like ancestors
NCA	Four	One-Human race	Four
NCA	God used big bang	God used big bang	Naturally from big bang
NCA	Idea is a myth	Each day is millions of yrs	They are a myth
NCA	~ 4.5 billion years	~ 4.5 billion years	~ 4.5 billion years
NCA	~ 12 and 20 billion years	~ 12 and 20 billion years	~ 12 and 20 billion years
NCA	extinct ~ 65M yrs ago	extinct ~ 65M yrs ago	Extinct ~ 65M yrs ago
NCA	Always a part of life	Always a part of life	Always a part of life
NCA	The Bible maybe	The Bible	No, personal preference
NCA	NCA	Inerrant as properly interpreted by experts	Just another book
NCA	Local flood in Middle East	Local flood in Middle East	Just a myth
NCA	Evidence for millions of years	Evidence for millions of years	Evidence for millions of years
NCA	Evidence for evolution	Evidence for millions of years of creation	Evidence for evolution
NCA	Myth	Modest vessel or myth	Myth

NCA = No consistent answer

Intentionally left blank

AMINO ACIDS

Amino Acid Table		
alanine (L-alanine)	Non-essential	produced by the body
arginine (L-arginine)	**Essential***	**not produced by the body adults suffering liver disease**
asparagine (L-asparagine)	Non-essential	It is produced by the body
aspartic acid (L-aspartic acid)	Non-essential	It is produced by the body
cysteine (L-cysteine)	Non-essential	It is produced by the body
glutamic acid (L-glutamic acid)	Non-essential	It is produced by the body
glutamine (L-glutamine)	Non-essential	It is produced by the body
glycine	Non-essential	It is produced by the body
histidine (L-histidine) -	**Essential***	**It is not produced by the body in children**
isoleucine (L-isoleucine) -	**Essential**	**It is not produced by the body**
leucine (L-leucine) -	**Essential**	**It is not produced by the body**
lysine (L-lysine) -	**Essential**	**It is not produced by the body**
methionine (L-methionine)	**Essential**	**It is not produced by the body**
phenylalanine (L-phenylalanine)	**Essential**	**It is not produced by the body**
proline (L-proline)	Non-essential	It is produced by the body
serine (L-serine)	Non-essential	It is produced by the body
taurine	Non-essential	It is produced by the body
threonine (L-threonine)	**Essential**	**It is not produced by the body**
tryptophan (L-tryptophan)	**Essential**	**It is not produced by the body**
tyrosine (L-tyrosine)	Non-essential	It is produced by the body
valine (L-valine)	**Essential**	**It is not produced by the body**

Abbreviation - Name	DNA Codon
ALA - alanine (L-alanine)	GCU or GCC or GCA or GCG
ARG - arginine (L-arginine)	AGA OR AGG OR CGA OR CGC OR CGG OR CGU
ASN - asparagine (L-asparagine)	AAU or AAC
ASP - aspartic acid (L-aspartic acid)	GAU or GAC
CYS - cysteine (L-cysteine)	UGU or UGC
GLU - glutamic acid (L-glutamic acid)	GAA or GAG
GLN - glutamine (L-glutamine)	CCA or CAG
GLY - glycine	GGC or GGA or GGG OR GGU
HIS - histidine (L-histidine) -	CAU or CAC
LLE - isoleucine (L-isoleucine) -	AUU or AUC or AUA
LEU - leucine (L-leucine) -	UUA or UUG or CUU or CUC or CUA or CUG
LYS - lysine (L-lysine) -	AAA or AAG
MET - methionine (L-methionine)	AUG
PHE - phenylalanine (L-phenylalanine)	UUU or UUC
PRO - proline (L-proline)	CCU or CCC or CCA or CCG
SER - serine (L-serine)	UCU OR AGC OR AGU OR UCA OR UCG OR UCU
TRE - threonine (L-threonine)	ACU or ACC or ACA or ACG
TRP - tryptophan (L-tryptophan)	UGG
TYR - tyrosine (L-tyrosine)	UAU or UAC
VAL - valine (L-valine)	GUU or GUC or GUA or GUG

UNSUBSTANTIATED CLAIMS OF
THE EVOLUTIONIST

B

Unsubstantiated Claims of The Evolutionist

Panspermia
Vestigial Organs & Structures
Tree of Life and Open Fraud
Unfounded Missing Links
Natural Selection?
Homology

Panspermia: an attempt to justify belief in evolution.

Panspermia, specifically the concept of "Directed Panspermia," suggests that life on Earth may have been intentionally sent here by intelligent beings from another planet. This idea, though unconventional, was seriously proposed by Francis Crick—the co-discoverer of the DNA structure—along with chemist Leslie Orgel in a 1973 article published in the journal *Icarus*. In their abstract, they described Directed Panspermia as the theory that microorganisms were deliberately transmitted to Earth by an advanced extraterrestrial civilization. Crick recognized that the intricate complexity of DNA could not be accounted for by natural processes alone, yet he was unwilling to acknowledge a divine Creator as the source of life. (Williams, Stephen. *What Your Atheist Professor Doesn't Know (But Should)*, 2013, p. 154)

Panspermia may seem a far-fetched concept, but there are already many articles on the Internet suggesting it because life on Earth is made of left-handed amino acids, almost exclusively. To illustrate my point, do an Internet search for left-handed amino acids. You will find an astounding number of articles that discuss meteors bringing these left-handed amino acids to Earth. Ronald Breslow, Ph.D. University Professor at Columbia University and former President of the American Chemical Society, described how the amino acid signature came from outer space.

> "With the exception of a few right-handed amino acid-based bacteria, left-handed "L-amino acids" dominate on earth. Amino acids delivered to Earth by meteorite bombardments left us with those left-handed protein units." *In a report at the 235th national meeting of the American Chemical Society*

Vestigial organs & Structures

German anatomist Robert Wiedersheim, in 1893, estimated that there were 86 such structures. He claimed that this proved that

Darwin's theory was correct because these organs and systems were no longer necessary and must have been left over from earlier transitions. He was quoted in the now-infamous Scopes Trial in 1925 by Horatio Hackett Newman, a zoologist at the University of Chicago, as proof of evolution. Just because they cannot find a reason for the existence of an organ or feature does not mean they never will. Some still claim that there is "junk" in the DNA molecule, but I think that DNA junk is quickly being disproved.

For our purposes here, we will only mention a few of the most well-known instances of supposed vestigial structures that persist today.

Traditionally, evolutionists have cited vestigial organs—such as the appendix, tonsils, coccyx (tailbone), and even body hair—as "leftovers" from an evolutionary past, supposedly rendered function-less over time. However, modern scientific discoveries have increasingly challenged this assumption, revealing that many of these so-called vestigial structures have critical roles in the body today.

1. Appendix

In the context of created biology, the existence of the appendix offers critical information about the harmonious and intricate design of the human body. Every organ and system works in conjunction with others, illustrating the complexity and predictability inherent in biological systems, as conceived by the Creator. It has been observed that the appendix serves as a "safe house" for useful bacteria, which can repopulate the intestine after environmental disruptions. This functionality echoes the principles of intelligent design—how each aspect of anatomy is expected to serve more than one purpose, contributing to the overall health and adaptability of the body. In involving the immune system and facilitating microbiota regulation, the appendix exemplifies a sophisticated biological characteristic adapted to improve human resilience against disease.

This design presents a teleological (the doctrine of design) perspective, postulating that the human body is not an accident of evolutionary processes, but a wonder of intentional design. Each component, including the appendix, incorporates a purposeful function that supports the body's vitality. Our understanding of the contributions of the appendix to digestive health fits the creationist affirmation that the human body was created with inherent wisdom, providing tools for survival and flourishing in a complex world. As scientific research continues to explore the multifaceted roles of the appendix, it is becoming increasingly evident that this organ is proof of the complexity of the design found in human anatomy, challenging the reductionist views that categorize it as evolutionary debris. The appendix is often rejected as a vestigial organ, but a more in-depth examination of its anatomical structure reveals a thoughtful conception indicating intentional creation. Anatomically, the appendix is a narrow tubular structure that extends from the cecum, the beginning of the large intestine. Its distinctive shape and positioning suggest a role beyond simple rudimentary function. One of its most notable characteristics is the composition of its wall, which is rich in lymphatic tissue. This abundance of lymphoid follicles reflects critical involvement in the immune response, attesting to the role of the appendix in the maintenance of intestinal health and overall physiological homeostasis.

2. Tonsils

Once routinely removed, tonsils are now understood to be Human tonsils are an integral component of the immune system, comprising a collective group of strategically located lymphoid tissues in the oral and pharyngeal regions. Anatomically, three distinct types of tonsils are recognized: palatine tonsils, pharyngeal amygdala (usually called adenoid), and lingual tonsils. Palatine tonsils, most commonly recognized due to their visibility during an oral examination, are situated bilaterally in the oropharynx, the region in the back of the throat. Each amygdala appears as an oval mass of lymphoid tissue, contributing to the first line of defense of the body against pathogens in the air and those ingested. The pharyngeal amygdala, located on the roof of the nasopharynx, is less visible but plays a crucial role in immune surveillance in the upper respiratory tract. Finally, the lingual tonsils reside at the base of the tongue, adding another layer of lymphoid tissue to combat pathogens entering the oral cavity.

The composition of tonsils as lymphoid tissue underlines their function within the lymphatic system. They are rich in lymphocytes, particularly T cells and B cells, which are critical actors in the adaptive immune response. As primary lymphoid organs, tonsils also facilitate antibody production and immune cell activation after exposure to antigens, making them vital within the larger immune structure of the body. The structure of tonsils, characterized by crypts and folds in their epithelium, increases their ability to arrest and process pathogens, subsequently stimulating innate and adaptive immunity. This project amplifies their functional role in the identification, neutralization, and memory of specific antigens, thus contributing to long-term immunity.

Despite widespread misconceptions, such as the notion that tonsils do not serve essential functions and are merely vestigial, a comprehensive understanding of their anatomy and function dissipates these myths. Some argue that, in cases of chronic tonsillitis or obstructive sleep apnea, the removal of tonsils is justified; however, it is essential to recognize that tonsils offer substantial immunological benefits, especially during childhood, when the immune system is still developing. Scientific literature, including Netter's (2022) discoveries, emphasizes that tonsil removal can predispose individuals to an increased risk of respiratory infections and a compromised immune response. Thus, it is imperative that health professionals and patients consider the functional implications of the anatomy of the amygdala by deliberating surgical interventions.

3. Coccyx (Tailbone)

The coccyx is the terminal segment of the vertebral column, articulating with the sacrum. Its unique curved shape and fusion of small bones provide a stable base for various anatomical attachments.

Function and Purpose

1. Support and Weight Bearing:

One of the primary functions of the coccyx is to provide a stable tripod-like support when a person is sitting. Along with the ischial tuberosities (the "sit bones" of the pelvis), the coccyx helps bear the weight of the upper body, especially when leaning back. This distribution of weight reduces pressure on other parts of the pelvis and spine, contributing to comfort and stability.

2. Attachment Point for Muscles and Ligaments:
The coccyx serves as a crucial attachment point for several important muscles, tendons, and ligaments of the pelvic floor. These include:

Gluteus Maximus: A major muscle responsible for hip extension and external rotation.

Levator Ani Muscle: A group of muscles that form the pelvic floor, supporting the pelvic organs (bladder, intestines, uterus/vagina in females) and playing a vital role in continence (control of bowel movements).

Anococcygeal Ligament: This ligament connects the coccyx to the anus, providing support for the anal canal.
These attachments are essential for maintaining proper posture, facilitating defecation, and providing stability to the pelvic region.

3. Shock Absorption:
While small, the coccyx can also contribute to shock absorption when sitting or during certain movements, helping to distribute forces and protect the lower spine.

From our perspective, the coccyx is a clear example of intelligent design. Its specific shape, location, and the array of muscles and ligaments that attach to it demonstrate a purposeful integration into the human musculoskeletal system. It's not a "leftover" but a foundational anchor for critical pelvic floor structures that are essential for:

Support and Stability: Providing a stable base for sitting and distributing body weight.

Organ Support: Holding pelvic organs in place.
Continence: Playing a key role in bowel control.

Movement: Facilitating the function of major leg and hip muscles. The idea that the coccyx is a "vestigial tail" is based on an evolutionary interpretation that doesn't fully account for its functional significance in the human body. Instead, its design points to a Creator who intricately fashioned each part of the human anatomy with specific, interconnected purposes.

In summary, the coccyx is a small but mighty bone, perfectly designed to provide support, anchor vital muscles, and contribute to the overall stability and function of the human pelvis.

4. Body Hair and Eyebrows

Human body hair, though often less dense and prominent than the fur of many animals, is not a random or vestigial feature. From a creation science viewpoint, every aspect of the human body is purposefully designed by a Creator, and body hair serves several important functions that contribute to our well-being and sensory experience.

Structure and Distribution

Body hair varies in thickness, length, and density across different parts of the body and among individuals. It grows from hair follicles embedded in the skin and is composed primarily of keratin.

Function and Purpose

1. Sensory Perception:

Perhaps one of the most significant roles of body hair is its sensory function. Each hair follicle is surrounded by nerve endings. Even the slightest movement of a hair, such as from a gentle breeze or an insect crawling on the skin, stimulates these nerves. This allows us to detect subtle changes in our environment, providing an early warning system for potential irritants, parasites, or even changes in airflow. This heightened tactile sensitivity is a sophisticated design feature.

2. Protection:
While not as thick as animal fur, human body hair still offers a degree of protection:

Scalp Hair: Protects the head from UV radiation from the sun, preventing sunburn and reducing heat absorption. It also provides insulation against cold.

Eyebrows and Eyelashes: Protect the eyes from sweat, dust, and debris, acting as a barrier and directing moisture away from the eyes. Eyelashes also trigger a blink reflex when touched, protecting the eye from foreign objects.

Nose Hair and Ear Hair: Filter out dust, pollen, and other airborne particles, preventing them from entering the respiratory system or ear canal.

Axillary (Armpit) and Pubic Hair: These areas are rich in sweat glands. Hair in these regions can help reduce friction during movement and may play a role in dispersing pheromones, though this is a more complex area of study.

3. Thermoregulation (Minor Role):
While not as effective as dense fur, body hair can still play a minor role in thermoregulation. When we get cold, tiny muscles attached to hair follicles (arrector pili muscles) contract, causing hairs to stand on end (goosebumps). This action, while more pronounced in animals with thick fur to trap a layer of insulating air, still provides a slight insulating effect in humans and is a remnant of a more robust thermoregulatory system.

4. Aesthetic and Social Functions:
Beyond purely biological functions, hair also plays a significant role in human appearance, identity, and social interaction. Its variation in color, texture, and style contributes to human diversity and personal expression, which can be seen as part of the Creator's design for individuality.

From a creation science viewpoint, body hair is not a "leftover" from a furry ancestor, but rather a testament to the Creator's intricate design. Each type of hair, in its specific location, serves a purpose:

The sensory function of hair, particularly its ability to detect minute stimuli, points to a finely tuned warning system.
The protective roles of hair on the scalp, eyes, nose, and ears demonstrate foresight in safeguarding vulnerable areas of the body.
Even the minor thermoregulatory role, though less critical in humans than in other mammals, shows a consistent design principle.
The presence and distribution of human body hair, therefore, are not random evolutionary accidents but are part of a holistic, intelligently designed system that contributes to our protection, sensory awareness, and overall well-being.

5. Goosebumps (Arrector Pili Muscles)
Though often deemed a useless evolutionary leftover from fur-covered ancestors, the reflex of goosebumps can still serve a purpose. It helps trap warmth by causing body hair to stand up—minimally effective in modern humans, but still part of a broader design system involving emotional response and thermoregulation.

6. The Thymus
The thymus is a part of the lymphatic system and is where T cells are called to action, without it, we would be subject to various infections with no way to fight them.

For over 2,000 years, the purpose of the thymus gland remained a mystery to medical professionals. In more recent history, it was dismissed by modern physicians—much like the appendix—as a useless, vestigial organ that had either lost its function or never had one. However, scientific advances in recent years have revealed that the thymus is far from useless. In fact, it plays a central role as the master gland responsible for regulating the body's complex immune system, which defends us against infectious diseases. Furthermore, recent experiments suggest that other organs once thought to be vestigial, such as the appendix, tonsils, and adenoids, may also contribute significantly to the body's antibody responses.

(*The Useless Gland that Guards Our Health, Reader's Digest, November 1966, pp. 229, 235*)

From our perspective, the presumption that certain organs are "vestigial" arises more from ignorance than evidence. Just because a structure's function was not initially understood does not mean it has no purpose. History repeatedly shows that labeling something "useless" often reflects our limited understanding—not a flaw in design. In fact, over time, many so-called vestigial structures have been vindicated by scientific research.

Rather than being evolutionary leftovers, these features are better understood as parts of a well-engineered biological system. The intricacies and redundancies found in the human body reveal a Creator who designed organisms with complexity, purpose, and resilience. Just as engineers incorporate backup systems and multipurpose components into machines, so too does God's design reflect forethought and intentionality, not randomness or decay.

In conclusion, vestigial structures do not point to an evolutionary past, but rather underscore the richness of a creation that was "very good" from the beginning (Genesis 1:31). Their functions, once misunderstood, now testify to the wisdom and foresight embedded in God's design of the human body.

Darwin's Tree of Life

Over a century after Ernst Haeckel's "tree of life" was published in the book *The Evolution of Man (1879), there are still no transitional fossils or any evidence that would substantiate the claim.*

For an extended period, the ambition to create a detailed "tree of life" has been viewed as the ultimate goal in evolutionary biology. Eric Bapteste, an evolutionary biologist affiliated with the Pierre and Marie Curie University in Paris, remarked that scientists were optimistic about nearing this milestone just a few years ago. Nevertheless, the emergence of conflicting evidence has since undermined this initiative, casting significant doubt on its credibility.

Currently, numerous biologists argue that the conventional tree model is antiquated and ought to be discarded. Bapteste has even asserted, "We have no evidence at all that the tree of life is a reality." This insight has prompted some researchers to suggest that a fundamental reevaluation of biology may be essential. *(Graham Lawton, "Why Darwin was wrong about the tree of life," New Scientist Magazine)*

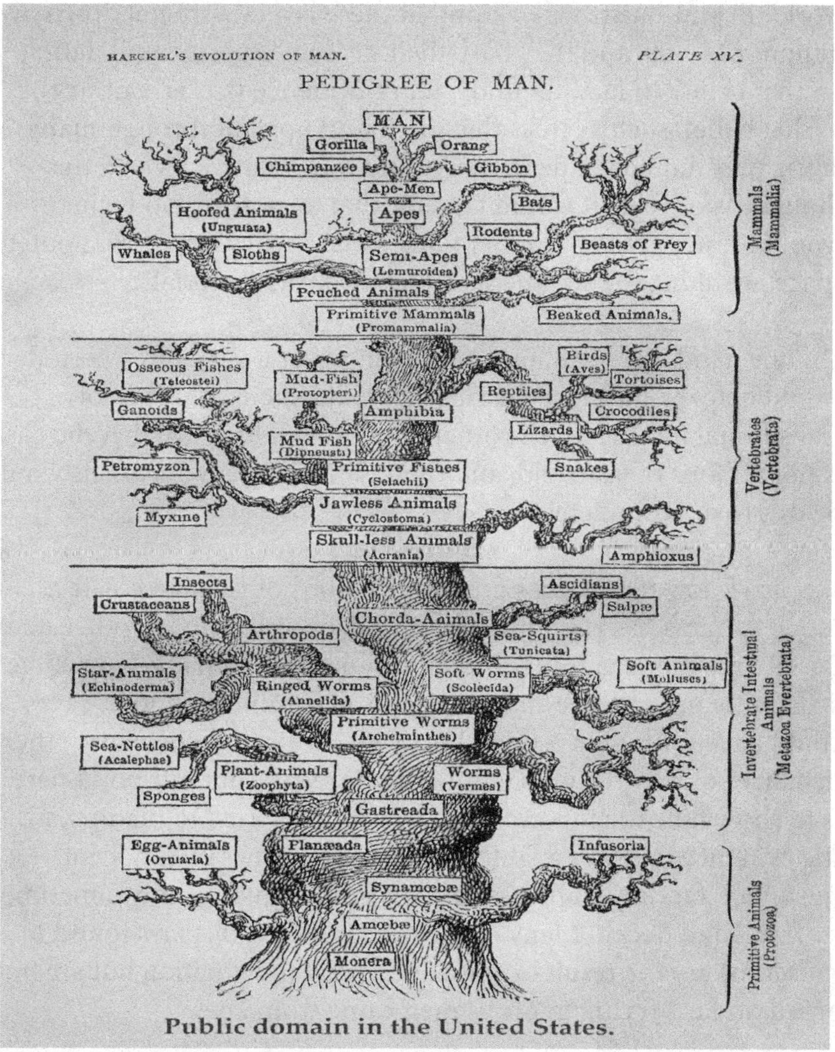

Public domain in the United States.

Deliberate Fraud or Mis-characterization:

It is my opinion that they will never find transitional fossils. That is a safe bet because after more than 150 years, none have been found. If it were true, we would find millions of them. The reason I hold that belief is that I also believe that God sent judgment on the earth in Noah's flood. There is overwhelming evidence of a global flood, but that is a subject for another book.

A global flood would lay down sediments as we find all over the world. Evolutionists will claim that these layers were put down over billions of years, and they base their geologic column and dating on that belief. In fact, we find a lot of evidence that is contrary to that belief, such as trees that are found upright through many layers of sediment. A dead tree would never have survived the thousands of years it would take for that to be true. And where would these layers of sediment originate? If they came from outer space, wouldn't the sediments all be the same material?

On one hand, a global flood happening over a very short period would help evolutionists because it would excuse the lack of transitional fossils. On the other hand, it would prove that the earth is not billions of years old, and their dating system for fossils would be disproved completely.

We also know that it does not take millions of years to create a fossil. A man's hat was found fossilized, as well as many other items. Based on a modern mineral water fountain, which created a huge stalagmite mound, the stalactites and stalagmites found in caves could have formed in a very short time. We could cite many other instances of why the global flood was a reality, but our point here is to show how the followers of Darwin have tried to manipulate the evidence to prove a faulty belief—sometimes by innocent misunderstanding, sometimes by wishful thinking, and sometimes by fraud and deceit. Many of the drawings and displays found in museums are the result of someone's vivid imagination but are not based on fact, trying to prove men came from apes.

Missing Links - Still missing

When you can't prove it, use your imagination

Piltdown man - Eoanthropus dawsoni which means 'Dawson's Dawn Man'.

A worker was digging in a gravel pit near Piltdown, in Sussex, in southern England, and found what appeared to be a skull. He turned it over to Dr. Charles Dawson. In 1912, Dawson found a jawbone. The intense desire of evolutionists to find a missing link made them easy prey for a hoax.

This hoax was uncovered by Dr. Kenneth Oakley in 1953 but had lasted for nearly forty years. He proved that the skull was human and the jaw was from an orangutan. The jaw had been tampered with to look more human.

> In 1923, Franz Weidenreich examined the remains and correctly reported that they consisted of a modern human cranium and an orangutan jaw with filed-down teeth. - MacRitchie, Finlay (2011), *Scientific Research as a Career, CRC Press, p. 30*

Java Man - Pithecanthropus alalus

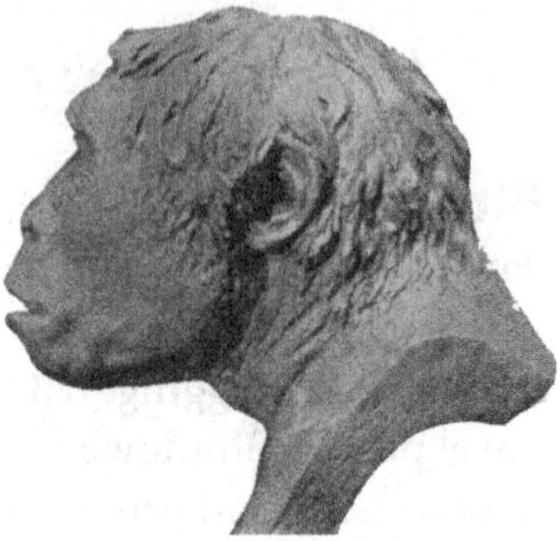

Java Man was found in 1891 by Eugene Dubois. Dubois was a former student of Ernst Haeckel. What they found was a part of the top of a skull, three teeth, and a thigh bone. Later, it was found that these bones were not discovered together but scattered more than 70 feet from each other. Haeckel named the find Pithecanthropus alalus (man without speech).

Above is an image they constructed to show what Java Man looked like at the time. That is wishful thinking, but it does not prove anything other than the creative mind of a biased artist wanting to depict an ape-like transition.

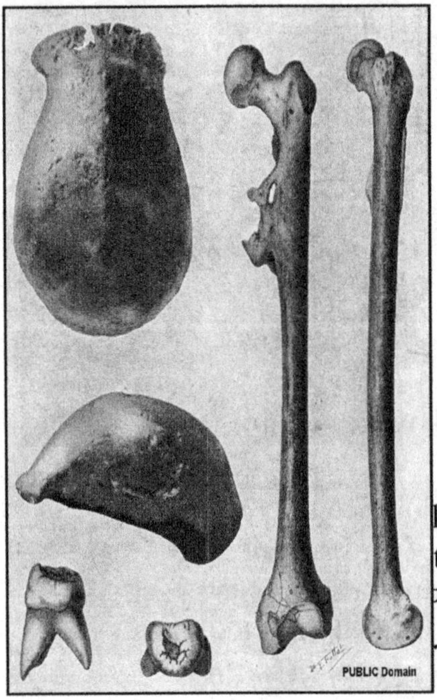

left are the bones with which they build the mythical Java. You can artist had been coached to show ape-like creature, even though the have no correlation to the image.

suspicions regarding the credibility of Dubois involve two distinctly human skullcaps that Dubois' expedition had uncovered He apparently failed to display human skullcaps when parading kept the skulls hidden under the ty years, then finally made them *'erloff, Tornado in a Junkyard: winism. Page 85.*

Nebraska Man - Hesperopithecus haroldcookii.

Harold Cook found Nebraska Man in 1922. This discovery occurred after Piltdown Man, which proved to be a fraud. Nebraska Man is based on a single tooth. The artist came up with the drawing above. Once again, the artist's rendition is completely false—very imaginative but completely fabricated in his mind. For over eighty years, it was promoted as an early ape-man. In the final analysis, the tooth was found to be from a pig, Prosthennops serus; this pig was found still alive in Paraguay in 1972.

> "I believe this is a case in which a scientist made a man out of a pig and the pig made a monkey out of the scientist." *Dr. Duane Gish*

Skeleton and restoration model of Neanderthal (La Ferrassie 1). Exhibit in the National Museum of Nature and Science, Tokyo, Japan.

Neanderthal man - Homo neanderthalensis

Discovered near Hamburg, Germany, by Professor Reiner Protsch.Protsch dated the find at 36,000 years ago, filling in the gap between modern man and this pre-human creature. However, in 2005, Thomas Terberger discovered the hoax, and Protsch resigned in disgrace when a Frankfurt University panel ruled he had "fabricated data and plagiarized the work of his colleagues."

A German anatomist, Rudolf Virchow, said the fossil was the remains of a modern man afflicted with rickets and osteoporosis. This is what gave the Neanderthal its distinctive but ape-like appearance.

Erik Trinkaus, a prominent paleoanthropologist and expert on Neandertal biology and human evolution, has been known to say that there is nothing in Neanderthal anatomy that conclusively indicates abilities inferior to those of modern humans.

> **"... you look and look again until you find what you know must be there."**
> - *Origins,* Richard Leakey and R. Lewin, p. 106, 1977.

In reconsidering his views on human evolution, David Pilbeam (*"Rearranging Our Family Tree,"* Human Nature, p. 45, 1978.) acknowledged a personal transformation in his approach to science.

He became more conscious of the implicit assumptions that often go unnoticed in scientific thinking and made a greater effort to identify and challenge them. Pilbeam recognized that many assumptions might only become evident over time, as current beliefs eventually reveal themselves to be misconceptions. He admitted that, particularly in the field of paleoanthropology, the scarcity of data allows theoretical frameworks to significantly shape interpretation. Historically, he noted, these theories have often mirrored prevailing ideologies more than the actual evidence.

Archaeopteryx

> Paleontologists have tried to turn Archaeopteryx into an earth-bound, feathered dinosaur. But it's not. It is a bird, a perching bird. And no amount of 'paleobabble' is going to change that. - Johnathan Sarfati "Archaeopteryx (unlike Archaeoraptor) is NOT a hoax—it is a true bird, not a 'missing link'." *Answers in Genesis, March 24, 2000.*

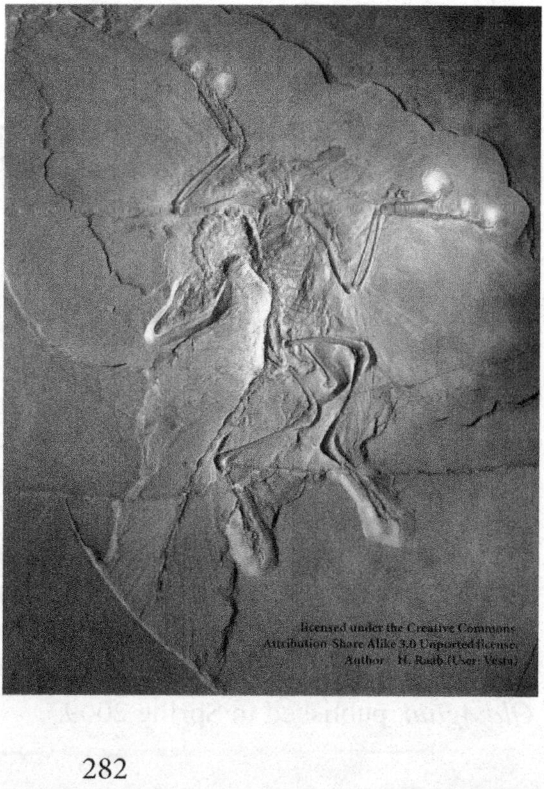

Sir Fred Hoyle, though an evolutionist, conducted a detailed examination of the two most prominent Archaeopteryx fossil specimens and concluded they were fraudulent. Upon studying both the front and back slabs, Hoyle and his team discovered inconsistencies between them, noting that the slabs did not properly align. They identified a specific alteration to the left wing, differing from its depiction in an 1863 illustration, and concluded that the feather impressions appeared to have been manually imprinted rather than naturally fossilized. Additionally, they found evidence that cement blobs had been used in the etching process. When the researchers requested access to advanced testing methods, including electron microscopy and carbon-14 dating, the museum denied the request and subsequently removed the specimens from further examination. These findings were reported in several publications, including the British Journal of Photography (March–June 1985), The New York Times (May 7, 1985), and New Scientist (issues 1447 and 1467).

Lucy: Australopithecus afarensis

Lucy is another example of wishful thinking and imagination. She was discovered by Donald Johanson in 1973 near Hadar in Ethiopia. The remains were named Lucy in tribute to the Beatles song, "Lucy in the Sky with Diamonds," which happened to be playing at the camp.

> **"The only certainty in this data-poor, imagination-rich, endlessly fascinating field is that there are plenty of surprises left to come."**
> **Michael Lemonick Time magazine**
> *"How Man Began" March 14, 1999*

Kevin Anderson references a Nova episode in which Dr. Owen Lovejoy discussed the anatomy of Lucy's hipbone, noting that it resembled that of a primate more than a human. Lovejoy assumed the discrepancy was due to post-mortem damage and proceeded to reshape a cast of the pelvic bone to reflect what he believed to be a more human-like structure. Anderson highlights the circular reasoning involved in this process—using evolutionary assumptions to guide the reconstruction, and then presenting the modified reconstruction as evidence supporting human evolution. This critique was outlined in Anderson's editorial, *What Is New Is Old Again,* published in Spring 2009.

Australopithecus Was a Well-Adapted Tree Climber. New research has confirmed that an extinct type of ape called Australopithecus afarensis was a well-adapted tree climber. *ICR Daily Science Updates, November 12, 2012.*

"The australopithecines known over the last several decades from Olduvai and Sterkfontein, Kromdraai and Makapansgat, are now irrevocably removed from a place in a group any closer to humans than to African apes and certainly from any place in a direct human lineage. All this should make us wonder about the unusual presentation of human evolution in introductory textbooks, in encyclopedias and in popular publications. In such volumes not only are australopithecines described as being of known bodily size and shape, but as possessing such abilities as bipedality and tool-using and -making and such developments as the use of fire and specific social structures. Even facial features are happily (and non-scietifically) reconstructed.". - Charles Oxnard *The Order of Man: A Biomathematical Anatomy of the Primates, p332.* (Former professor of anatomy at the University of Southern California Medical School)

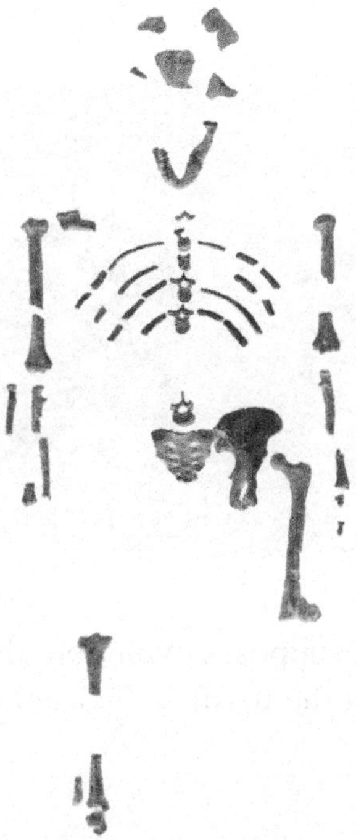

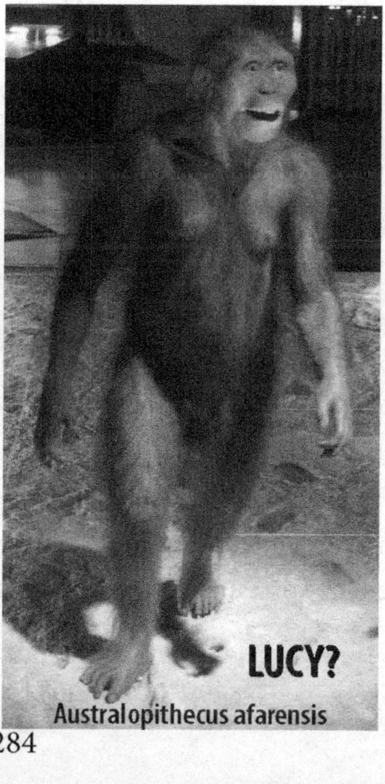

Ida - or Darwinius masillae (in honor of Charles Darwin).

I am indebted to Carl Kirby, of rforh.com Reasons for Hope - for pointing out the flaws about IDA in his DVD *Evoltuion: is that all ya got?*

However, it looks identical to modern lemurs. The movie and the website make the statement: "her anatomy has remarkable similarities to our own." The author makes the false claim that IDA "has a flat face like humans." Is he trying to use the power of suggestion to convince people? Personally, I don't see any similarity. Overlaying a lemur skull on IDA makes a perfect match.

The fossil was claimed to be a missing link. They came out with a book and a movie. You can see for yourself at: *Http://www. revealingthelink.com.*

The movie compares IDA to a lemur, its body structure, arms, legs, tail, and hands. How is it that they can then go on to say that this is the missing link? The movie states that the hands are like humans. But when compared to a lemur, they are similar. The comparison is then made to humans in that IDA has forward-facing eyes. A lemur has forward-facing eyes. IDA is said to be 47 million years old, but she looks similar to lemurs today.

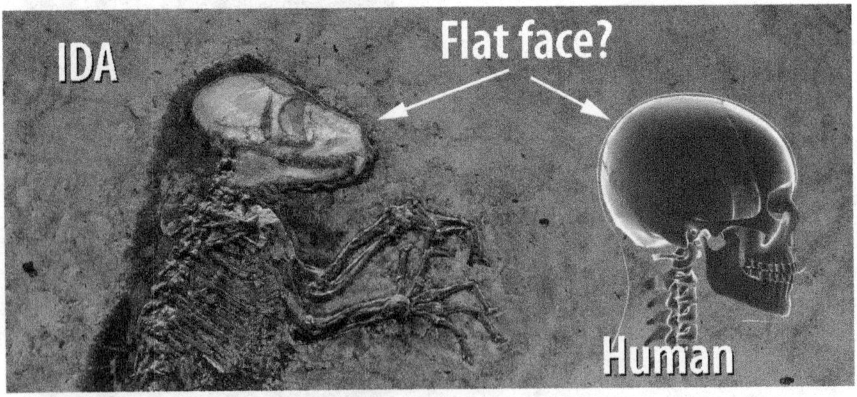

Fossils show "evolution" only if one presupposes evolution, then uses that presupposed belief to interpret the fossil. - *Answers in Genesis*

Once again, if there were transitions between IDA and humans, we would find massive evidence in the fossil record that shows a progressive transition from one form to another; there are none. In the final analysis, IDA is only an extinct lemur-like creature buried and preserved as a fossil.

> "the scientific study did not provide sufficient proof to consider Ida as a human ancestor, or even a primate ancestor. It has fingernails, something we have, as do most primates … but they've cherry-picked particular characteristics, and they've been criticized [by other scientists] for doing that." "It's not a missing link, it's not even a terribly close relative to monkeys, apes and humans, which is the point they're trying to make," says Carnegie Museum of Natural History curator of vertebrate paleontology Chris Beard. Quotes are from Robert MorleyColumnist A Multi-million-Dollar Monkey Fraud June 2, 2009 | From theTrumpet.com

Haeckel's Embryos

Romanes' 1892 copy of Ernst Haeckel's fraudulent embryo drawings

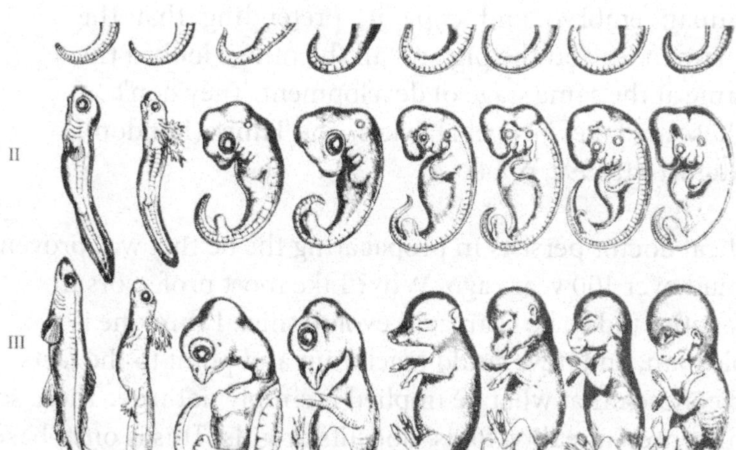

In doing research for this chapter, I came across a lecture series by a medical doctor preparing medical students to understand the major systems of the human body. The series is quite fascinating. The doctor explains the root meanings of the words used in medicine, making it much easier to understand. I was enjoying the lectures immensely... Then, out of the blue, he began describing the developing embryo. The doctor said, "Human development goes through the same evolutionary stages as other creatures go through, including gills." I knew immediately where this idea originated. It came from Haeckel's famous and fraudulent drawings. Haeckel deliberately drew the developing embryos to show that "ontogeny recapitulates phylogeny" (i.e., development replays evolutionary history). Wilhelm His (1831-1904) exposed the hoax in the German publication Unsere Koperform. Haeckel was discredited and banned from the ranks of trustworthy research scientists. However, these drawings were, and maybe still are, being used in some textbooks as proof for evolution. In any case, this doctor is still promoting the lie.

> "This is one of the worst cases of scientific fraud. It's shocking to find that somebody once thought to be a great scientist was deliberately misleading. It makes me angry … What he [Haeckel] did was to take a human embryo and copy it, pretending that the salamander and the pig and all the others looked the same at the same stage of development. They don't … These are fakes." Nigel Hawkes, The Times (London), August 11, 1997, p. 14).

This medical doctor persists in propagating the lie that was proven to be a fraud over 100 years ago. Why? Like most professors in our universities today, he is an avid evolutionist. I think he is a reasonable man, and if we could reach him and point to the facts, he would concede that what he implied is simply not true. These so-called "gills" are, in reality, folds, not slits or gills. This is one phase of the developing embryo, which eventually becomes specialized structures in the head and neck. Previously, we have alluded to the fact that there appears to be a function in the development of the embryo that determines the living system's body type during development. This is not fully understood currently; however, the DNA still has the parts list for building the proteins of the entity. But, in no way does it go through a recreation of the evolutionary past. Michael Denton observed that development through the earliest embryonic stages was completely different in amphibians, reptiles, and mammals.

However, throughout the doctor's lecture, whenever a unique function was introduced, he would say, "see what evolution has given us." Like all evolutionary sycophants, he never explained how the function would have evolved or how the system had survived without this function; he only gave blanket credit to evolution. It is no wonder, then, when we send innocent, God-fearing children to these universities, they end up rejecting God for Godless humanism. A continuous diet of evolutionary propaganda will take its toll, and blind allegiance to evolution becomes their way of life. Unless they look at the evidence critically, as many highly educated scientists have done in the past, they will forever be stuck believing a lie. This doctor illustrates how powerful this indoctrination is and how pervasive it is in all sciences and educational institutions, even so-called Christian universities. Some scientists have discovered the fallacy of Darwinism by examining the facts and have determined that evolution, as the neo-Darwinist explained it, is simply not possible. Because there is so much invested in the idea of evolution, it is unlikely that these proponents of neo-Darwinism will be shown the door as further proof of design is realized.

Natural Selection

Natural selection takes place in the phenotype (outward expression of an entity). For a creature to change, it must change at the gene level, and that change must be passed on to offspring.

The core issue is that natural selection operates at the phenotypic level—on the observable traits of organisms—not at the genetic level where mutations actually occur. Physicochemical forces and environmental pressures do not determine which nucleotide will be added next to a growing biopolymer chain. Mutations arise randomly at the genetic level, while environmental selection acts only later, at the level of the biopolymer's folded, functional structure. This selection occurs only after the biopolymer has already exhibited a specific algorithmic function, meaning that natural selection cannot guide the initial genetic changes that produce functional molecules. This distinction is emphasized in T.G. Dobzhansky's - *Discussion of Synthesis of Nucleosides and Polynucleosides with metaphoric Esters*

Finches Beaks:

Consider the variety of beaks in Darwin's Galápagos finches. God, in His infinite wisdom and design, has built a code so complex that it can produce an almost limitless variety of phenotypic expressions. However, offspring are always within a type and never outside that type. The change in finch beaks did not prove evolution; it only showed that under certain conditions (weather and rainfall), beaks necessary for cracking open seeds required a certain type of beak. Therefore, more birds with that type of beak would be born, and others would begin to die out. However, when the weather changes again, the other birds with different beaks begin to multiply once more.

What this proves is that your worldview (i.e., all things come from nature) will dictate how you interpret data. To the creationist, it is obvious that no new creature evolved, just a swing in the population of the finch type.

Peppered Moths:

The pictures used to prove natural selection were staged to support the theory of natural selection. Since there were more light-colored peppered moths in the early 19th century before industrial pollution became a major problem, Bernard Kettlewell, a British scientist, put forth the idea that since the tree trunks on which the peppered moths land had become dark with industrial pollution, the lighter ones would be more visible to predatory birds. The light ones would be eaten first. This would be proof of natural selection and, therefore, evolution.

This theory had problems; peppered moths do not land on the trunks of trees, and the photos that were taken initially were staged to make the theory appear true. In 1980, Kauri Mikkola, a Finnish zoologist, observed that the moths rest high up in the tree. So, the theory of industrial pollution proving natural selection in peppered moths is incorrect.

However, even if it were true, changes in a species from one color to another do not prove evolution; it only shows that the outward physical expression is controlled by the DNA of that kind. No new creature was created here, only a variation within a kind. I agree with that; we see it all the time—look at the variety of dogs. Just like the variety of beaks in Darwin's Galápagos finches, no evolution has taken place here.

The problem that we have today is that this fraud is still being taught in schools as proof of natural selection and evolution.

1. Geospiza magnirostris. 2. Geospiza fortis.
3. Geospiza parvula. 4. Certhidea olivacea.

Homology

Evolutionists use similarities to prove their ideas about how life evolved. Once again, the worldview of the Darwinists comes to the surface. Because they hold the view that only natural processes can explain life, they adamantly maintain that belief; because something is similar in shape or function, that means they came from the same lineage. This is revealed in the following quote from the Encyclopedia of Evolution.

> "Homo means 'the same.' The seven bones in the human neck correspond with the same seven, much larger, neck bones in the giraffe: They are homologues. The number of cervical vertebrae is a trait shared by creatures descended from a common ancestor. Related species share corresponding structures, though they may be modified in various ways." *The Encyclopedia of Evolution: Humanity's Search for Its Origins* - Richard Milner

For those of us who believe in a Creator, it is obvious that similarities prove that we all have the same designer. If something is well-designed, there is no need to make it different for the sake of being different.

> "There is not a trace at a molecular level of the traditional evolutionary series: fish to amphibian to reptile to mammal. incredibly man is closer to lamprey than are fish!" Michael Denton - *Evolution a Theory in Crisis*

> "By this we have also proved that a morphological similarity between organisms cannot be used as prod of a phylogenetic [evolutionary] relationship . . It is unscientific to maintain that the morphology may be used to prove relationships and evolution of the higher categories of units."
> —Herbert Nilsson, *Synthetische Artbildung*

"Techniques now exist for measuring the degree of similarity between most forms of life. These "genetic distances" are calculated by taking a specific protein and examining the sequence of its components. The fewer changes required to convert a protein of one organism into the corresponding protein of another organism, the closer their relationship. Similar comparisons can now be made between the genetic malarial of different organisms. The results of these studies seriously contradict the theory of evolution. There is not a trace of evidence on the molecular level for the traditional evolutionary series: simple sea life > fish> amphibians > reptiles> mammals. In general, each of the many categories of organisms appear to be equally isolated. One computer-based study, using cytochrome c, a protein used in energy production, compared 47 different forms of life. If evolution had actually occurred, this study should have found that, for example, the rattlesnake was most closely related to other reptiles. Instead, based on this one protein, the rattlesnake was most similar to man. Hundreds of similar contradictions have been discovered." —Walter Brown, *In the Beginning*

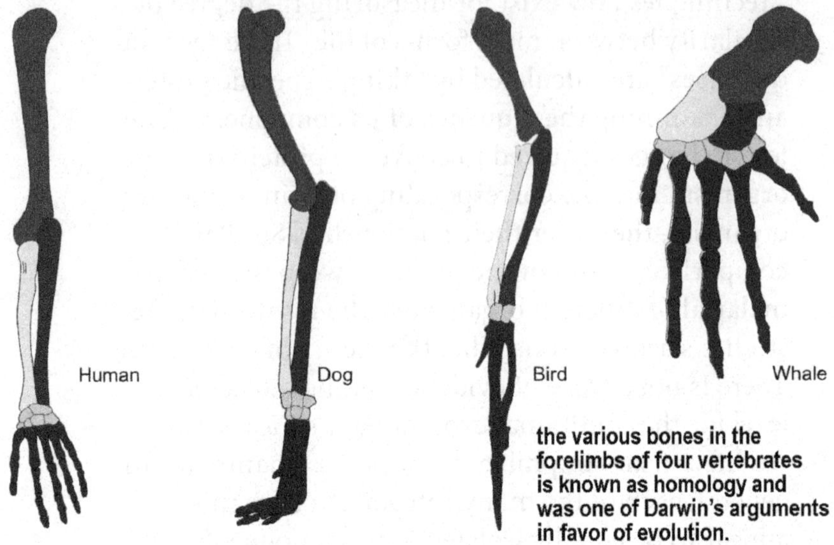

There are many other examples we could give. This chapter is only an overview of the many flaws in evolutionary thinking. I suggest a great book on the subject, if you were interested in knowing much more about unsubstantiated claims made by neo-Darwinists called *"Icons of Evolution - Why much of what we teach about evolution is wrong" by Jonathan Wells.*

Searching For Truth

GLOSSARY

C

Glossary of Terms used in the book

Abiogenesis: life arising from non-life, This idea has never been proven and is chemically impossible.

Acetyl tags: the presence of acetyl molecules will cause DNA to be not so tightly wound on a histone allowing the gene to be easily read during transcription. Both the acetyl and methyl tags are passed on during cell division and can be passed on to offspring.

Active Transport: The process where molecules enter or exit the cell but require energy (ATP) to work.

ADH (Anti-Diuretic Hormone)
A hormone produced in the hypothalamus and secreted by the pituitary to regulate water levels in the blood.

Adipocyte: A fat cell. Adipose tissue is crucial in energy storage and metabolic homeostasis.

adipogenesis: the production of fat or depositing of fat in the cell. It is also the conversion of carbohydrates or protein to fat.

Apoptosis: Programmed cell death, a process where cells are deliberately killed off.

Adenosine triphosphate (ATP): ATP is where the cell gets its energy for a multiple of different processes. When energy is spent it becomes Adenosine diphosphate. ATP is created during cellular respiration. It is said that you use about your body weight in ATP every day. The poison, cyanide, will kill you in 30 seconds because it completely shuts down ATP production in the cell.

> This motor is an excellent example of irreducible complexity because it is necessary in its entirety in order for even the simplest form of life to survive.
> Jerry Bergman, Ph.D.

Cellular respiration: this is the process whereby cells release energy from food molecules. This energy is critical for life to exist.

Chromatin: Is the material that makes up a chromosome.

Chromosome: one of the elements in the human cell nucleus that make up the genome. In humans, there are 46; 23 from mom and 23 from dad.

Codon: is a three letter nucleotide in DNA that codes for a specific amino acid. (see amino acids Appendix A)

Cytoplasm: is the region inside a cell but outside of the nucleus.

Cytosol: the liquid found inside cells.

Diffusion: a process where molecules enter or exit the cell without the use of energy (ATP)

DHAP: Dihydroxyacetone phosphate involved in many metabolic pathways, including glycolysis.

DNA Nucleotides: For our discussion in this book, a nucleotide is made up of a Nitrogenous base (like A,T, G, C or U) and has a 5 carbon sugar and a phosphate. This is one side of the DNA helix.

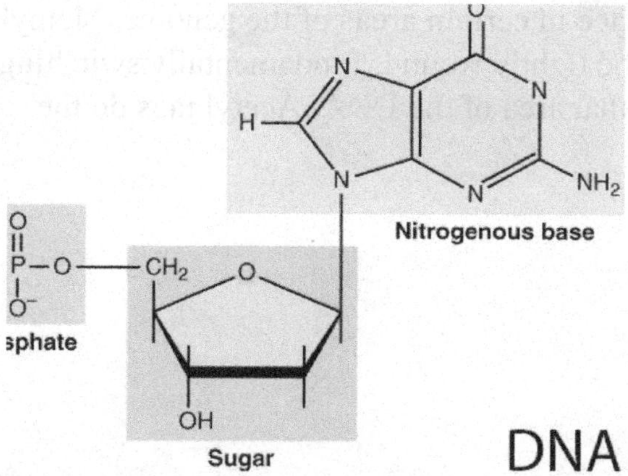

DNA Nucleotides

By OpenStax College

Electron Transport Chain: The third process in a series of chemical reactions which create energy for the cell (ATP)

Enzymes: is a catalyst to induce rapid changes in other substances, while remaining unchanged itself. Typically, you can tell if a protein is an enzyme if it has an "ase" ending like transfer***ase***.

Estrogenic: have a result like that of estrogen.

GADP: Acronym for glyceraldehyde phosphate, active in metabolic pathways.

Genotype: is the combination of the 23 chromosomes from mother and father, it is your DNA.

Glycolysis: "splitting sugars."

Golgi body: An organelle in the cell which is responsible for folding a protein after the process of translation into a polypeptide chain in the ribosome. The protein is then released to be used in the cell or outside the cell.

Histone: A protein around which DNA is wrapped is called a nucleosome. By wrapping DNA around histones, the double helix takes up less space, allowing it to fit in the nucleolus. Histones are also responsible for gene regulation, allowing or preventing transcription to take place in certain areas of the genome. Methyl tags cause histones to be tightly wound, fundamentally switching off gene expression in that area of the DNA. Acetyl tags do the opposite.

Helicase Enzymes:

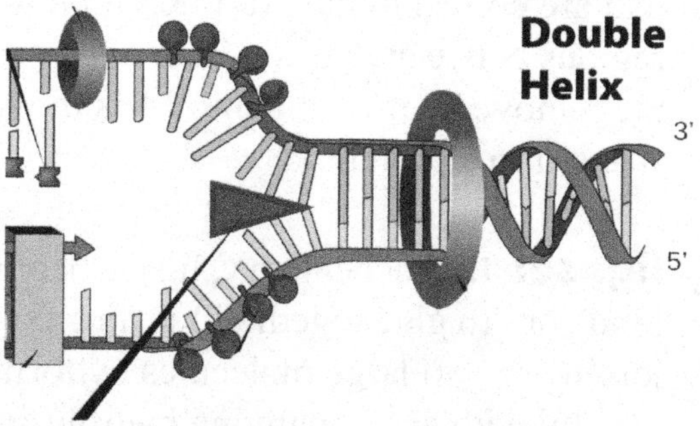

Enzymes speed up a process; otherwise, no work would get done in the time needed. The helicase enzymes are special enzymes that effectively "unzips" the DNA double helix to make the process of transcription possible.
Some of the functions of these enzymes are DNA replication, transcription, translation, recombination, and DNA repair.

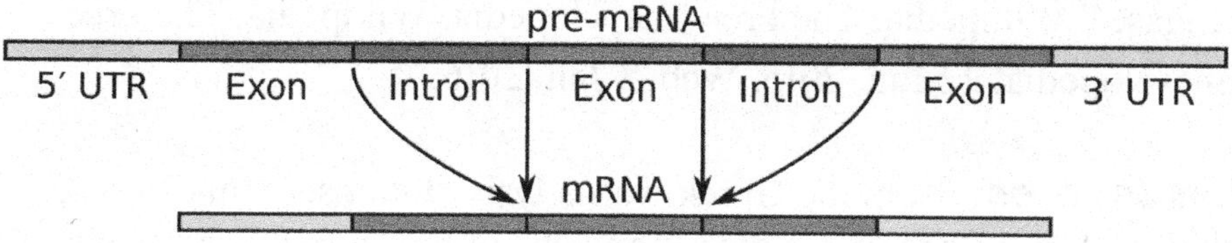

Public Domain

Hypothalamus

A gland near the Pituitary gland in the brain which is responsible for the regulation of body temperature and water levels.

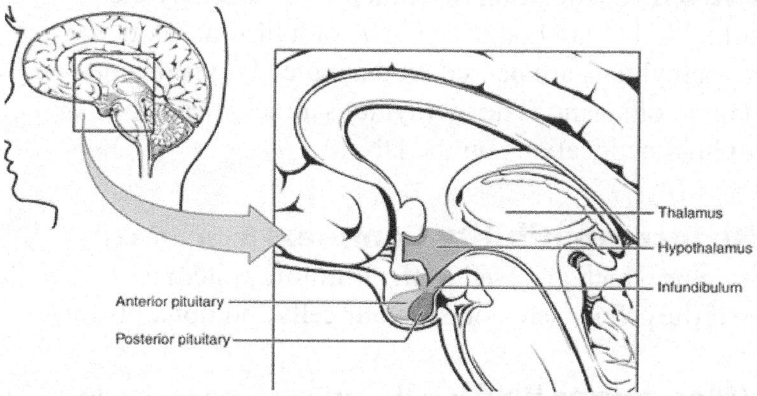

Introns (Intragentic Regions):
When reading the DNA string to make a protein, we find regions that are copied to the pre-mRNA known as introns. Introns are regions between exons within a transcribed mRNA string which are removed before the mRNA is moved to translation phase of the string into a protein.

ligase: In biochemistry, ligase (from the Latin verb ligāre — "to bind" or "to glue together") is an enzyme that can catalyze the joining of two large molecules by forming a new chemical bond, usually with accompanying hydrolysis of a small pendant chemical group on one of the larger molecules or the enzyme catalyzing the linking together of two compounds, e.g. enzymes that catalyze joining of C-O, C-S, C-N, etc. - Wikipedia contributors. "Ligase." Wikipedia, The Free Encyclopedia. Wikipedia, The Free Encyclopedia, 22 Jun. 2015. Web. 3 Jan. 2016.

Lysosomes: from the Greek words lysis "to loosen", and soma, "body". It is referred to as the ,"stomach" of the cell. It is a vesicle with the function of digesting bacteria and waste. It contains a wide variety of enzymes to accomplish its task. Some refer to it as, "suicide sacs" because it will merge with phagosomes to self-digest the cell effectively killing it.

Methyl Tags: gene reading machinery is blocked by the presence of a Methyl tag bound on DNA or a histone. Both the acetyl and methyl tags are passed on during cell division and can be passed on to offspring. The methyl tags attach themselves on Cytosine/ Guanine locations on the DNA

Major Histocompatibility Complex: these are cell surface proteins which are used in the immune system to determine if the cell is "self", one of your cells and not an invader.

mRNA (Messenger RNA): is the string of codons made in the transcription of DNA and after the introns are removed. mRNA will be moved to ribosomes outside of the nucleus, to be translated into a protein by linking amino acids.

Microtubules; form the "skeleton" of the cell. It is also used as route on which the organelles and materials can travel across the cell. During the mitosis phase of the cell cycle they form spindles, which are used to separate and manipulate chromosomes.

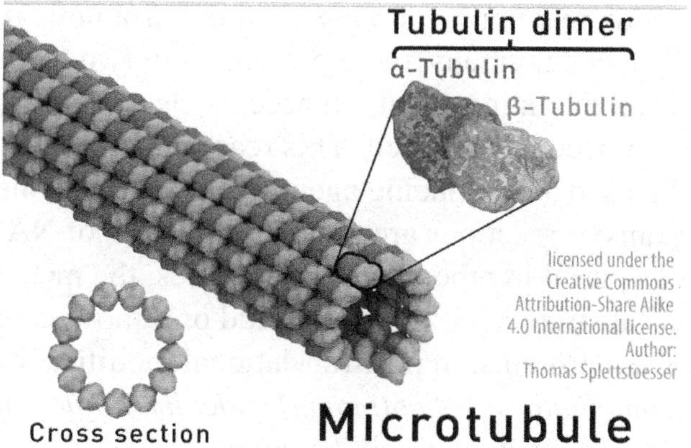

Mitochondrial DNA: is DNA found in the mitochondria. It is typically inherited from the mother. Mitochondrial DNA is a circular closed double helix with approximately 17,000 base pares

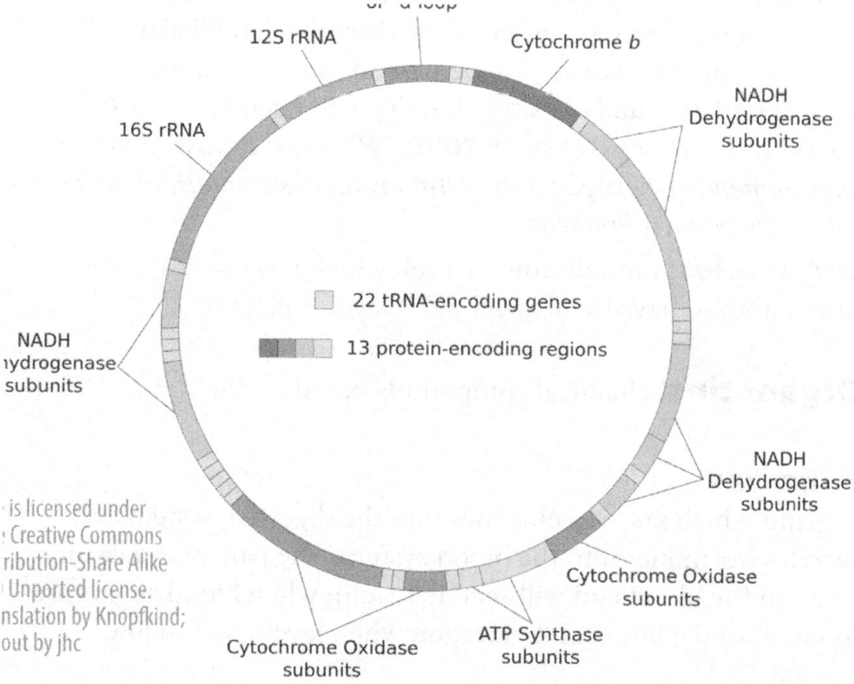

as compared to DNA in the cell which has three billion.

NADH, NAD+: Nicotinamide adenine dinucleotide - In metabolism, nicotinamide adenine dinucleotide is involved in redox reactions, carrying electrons from one reaction to another. The coenzyme is, therefore, found in two forms in cells: NAD+ is an oxidizing agent – it accepts electrons from other molecules and becomes reduced. This reaction forms NADH, which can then be used as a reducing agent to donate electrons. These electron transfer reactions are the main function of NAD. However, it is also used in other cellular processes, the most notable one being a substrate of enzymes that add or remove chemical groups from proteins, in posttranslational modifications. - *Wikipedia contributors. "Nicotinamide adenine dinucleotide." Wikipedia, The Free Encyclopedia. Wikipedia,*

Nucleosome: a single histone spool with its associated DNA. A sub-unit of chromatin.

Obesogens: Obesogens may be functionally defined as chemicals that inappropriately alter lipid homeostasis and fat storage, change metabolic setpoints, disrupt energy balance or modify the regulation of appetite and satiety to promote fat accumulation and obesity. - Kirchner S, Kieu T, Chow C, Casey S, Blumberg B (March 2010). "*Prenatal exposure to the environmental obesogen tributyltin predisposes multipotent stem cells to become adipocytes*"

Organelles: are subunits in a cell which have very specific functions and have their own lipid bilayer typically.

Organotins: chemical compounds based on tin

Pancreas

A gland which secretes enzymes into the digestive system and secretes hormones into the blood system. The pancreas measure levels in the blood and will secrete insulin when blood sugars are too high in the blood and glucagon when levels get too low.

Passive Transport: the movement of molecules across the cell

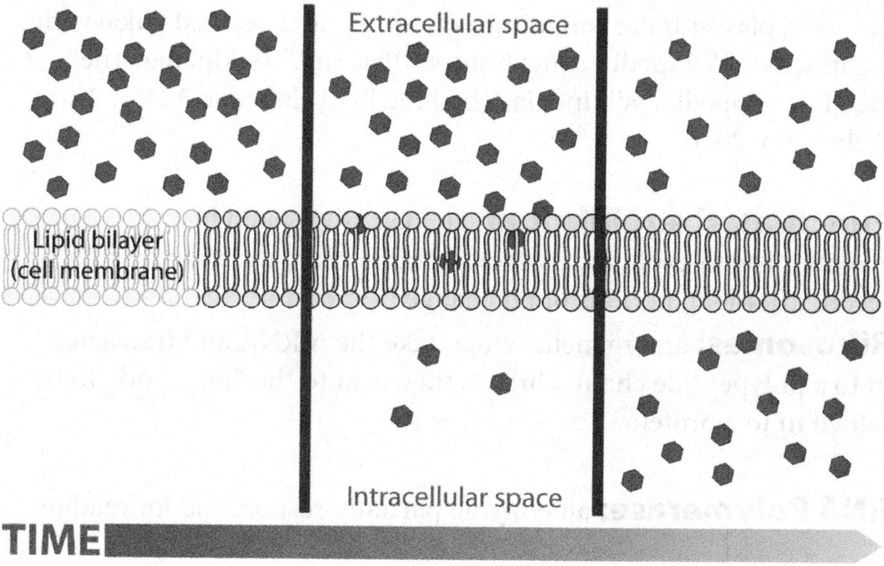

released into the public domain by its author, LadyofHats.

membrane without needing energy.

Pathogens: any microorganism that causes disease like: germs, viruses and other substances.

Phenotype: is the outward expression of your genotype, your skin color, the color of your hair among many other factors.

Phosphocycerate Kinase: is an enzyme that catalyzes the reversible transfer of a phosphate group from 1,3-bisphosphoglycerate (1,3-BPG) to ADP producing 3-phosphoglycerate (3-PG) and ATP. Like all kinases it is a transferase. PGK is a major enzyme used in glycolysis, in the first ATP-generating step of the glycolytic pathway. - *Wikipedia contributors. "Phosphoglycerate kinase." Wikipedia, The Free Encyclopedia. Wikipedia,*

Plasmids: A plasmid is a small DNA molecule within a cell that is physically separated from a chromosomal DNA and can replicate independently. They are most commonly found in bacteria as small, circular, double-stranded DNA molecules;

however, plasmids are sometimes present in archaea and eukaryotic organisms. - Wikipedia contributors. "Plasmid." Wikipedia, The Free Encyclopedia. Wikipedia, The Free Encyclopedia, 3 Nov. 2015. Web. 3 Jan. 2016.

Punctuated equilibrium: an idea propagated by evolutionists to explain why there is rapid changes in the fossil record after long periods of no change.

Ribosomes: an organelle which take the mRNA and translate it in to a polypeptide chain which is then sent to the Golgi body to be folded in to a protein.

RNA Polymerase: an enzyme partially responsible for reading

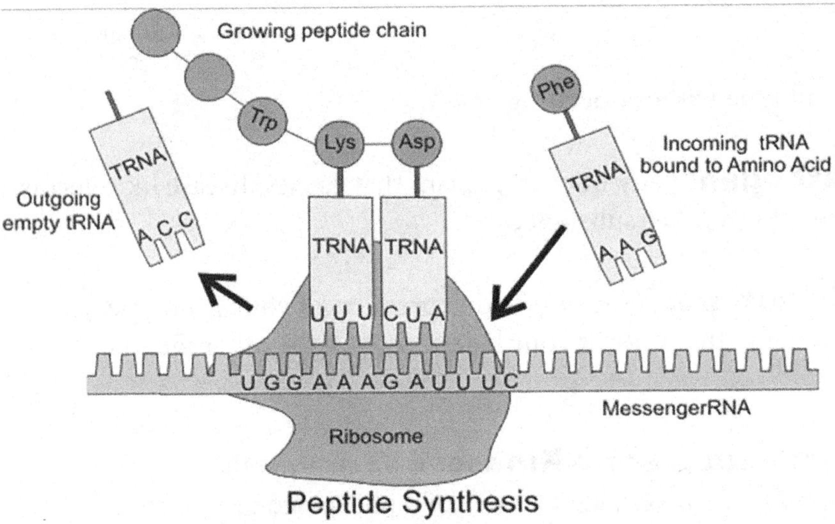

the gene for transcription. This transcripted string will then be translated into a specific protein.

Saltatory conduction: occurs in a break in the myelin sheath on a neuron axion, called the nodes of Ranvier, increasing the conduction velocity of action potentials to keep the signal moving across the axion. Otherwise, if the length of the sheath were too long, the loss of sodium traveling through the axion would cause the signal to weaken.

Transposon: A transposable element (TE or transposon) is a DNA sequence that can change its position within the genome, sometimes creating or reversing mutations and altering the cell's genome size. Transposition often results in duplication of the TE.

tRNA: (transfer RNA) tRNA brings a specific amino acid to the translation process which matches the codon on the mRNA string. This process connects the amino acids in a polypeptide chain which will ultimately be folded into a protein.

Transcription factors: a protein that binds to specific DNA sequences, thereby controlling the rate of transcription of genetic information from DNA to messenger RNA

Vestigial organs: organs that are believed to be left over from an evolutionary past and are no longer needed by the body.

Xenobiotic: is a foreign chemical substance found within an organism that is not normally, naturally produced by, or expected to be present within that organism. It can also cover substances which are present in much higher concentrations than usual. Specifically, drugs such as antibiotics are xenobiotics in humans because the human body does not produce them itself, nor are they part of a normal food.

SPECIAL SUBJECTS

D

Special Subjects

How the Autonomic Nervous System Keeps Us in Harmony and Handles Stress

The autonomic nervous system (ANS) is a vital component of the human nervous system that plays an important role in regulating involuntary physiological functions. These functions include heart rate, blood pressure, breathing, digestion and sexual excitement, which are essential to maintain homeostasis, the body's internal stability. Particularly important for adolescents, ANS adapts to changes in the environment and responds to stressful factors, which makes it a critical issue to understand general health during this development stage.

The ANS is divided into three main components: the ***sympathetic*** nervous system, the ***parasympathetic*** nervous system and the ***enteric*** nervous system. Each component has different functions that contribute to the body's ability to deal with stress and maintain balance. The sympathetic nervous system is often described as the "fight or flight" system. Prepare the body to respond to the threats perceived by increasing the heart rate, dilating the respiratory tract and redistributing blood flow to essential organs and muscles. This answer is crucial in times of stress, either due to academic pressures, social situations or physical threats.

In contrast, the parasympathetic nervous system is known as the "rest and digested" system. It retains energy and promotes relaxation by decelerating heart rate, stimulating digestion and facilitating body processes that are active during low-stress periods. These two systems work antagonistically; When one activates, the other is suppressed, ensuring that the body can respond properly to several situations. It is this balance between sympathetic and parasympathetic systems that allows homeostasis (Faulkner, Hathaway and Tolley 2003).

The enteric nervous system, often known as the "second brain", is a complex network of neurons that governs gastrointestinal function. Although sometimes it is considered separately from the ANS, it communicates with the central nervous system and the other parts of the ANS, illustrating the interconnection of body functions. This system plays a crucial role in the digestion and absorption of nutrients, which are important for adolescents who experience rapid physical growth (Harteveld et al. 2021).

Homeostasis is a state of equilibrium that the body tries to maintain despite external and internal changes. This balance is critical for survival and optimal functioning. The ANS helps maintain homeostasis regulating various processes, including body temperature, fluid balance and blood pressure. For example, when a person is in a hot environment, the body must cool to maintain a stable internal temperature. The sympathetic nervous system triggers sweat production and increases blood flow to the skin to facilitate heat dissipation. At the same time, the parasympathetic nervous system can help control hydration levels influencing the function of the kidneys (Tonhajzerová et al. 2002).

The importance of ANS extends beyond merely physical functions; It also plays an important role in emotional regulation and mental well-being. We often face several stressful factors, including academic challenges, social interactions and identity formation, which can lead to greater activation of the sympathetic nervous system. The research indicates that the ANS reacts differently from stress in adolescents compared to adults, showing a more pronounced sympathetic response (from Zambotti et al. 2018). This greater sympathetic reactivity can contribute to problems such as anxiety and mood disorders if they are not handled properly.

In addition, the relationship between ANS and stress is particularly critical during adolescence, a development phase characterized by significant changes in the body and brain. Studies have shown that cardiovascular autonomous functioning may differ depending on sex and age, suggesting that the maturation of ANS can influence adolescents faced with stress (Harteveld et al. 2021). As young people mature, their ability to regulate the balance between sympathetic and parasympathetic responses can also evolve, affecting their general health resistance and resistance (Faulkner, Hathaway and Tolley 2003).

A crucial appearance of ANS is its ability to adapt to chronic stress. Chronic activation of the sympathetic nervous system can lead to various health problems, such as hypertension, reduction of immune function and mental health disorders. On the contrary, a parasympathetic system that works well can cushion against the effects of stress, providing a general health protection mechanism. Consequently, participating in practices that promote parasympathetic activation, such as full attention, deep breathing and regular physical activity, can be especially beneficial for adolescents in stress management (from Zambotti et al. 2018).

Maintaining a balance between sympathetic and parasympathetic systems is essential for adolescents, particularly as they navigate the complexities of life during this significant development phase. The impact of lifestyle elections, such as diet, exercise, sleep and social interactions, in the functioning of ANS can be deep. For example, regular physical activity improves cardiovascular regional regulation, which leads to an improvement in heart health and stress resistance (Tonhajzerová et al. 2002).

We must also be educated about the importance of sleep and relaxation techniques to handle stress and promote better autonomous functioning. The poor sleep patterns can negatively influence the ANS, which leads to greater sympathetic activity and decreased parasympathetic activity during the day, which essentially interrupting the necessary balance for optimal health (Faulkner, Hathaway and Tolley 2003). Understanding the role of ANS can train us to make healthier decisions that promote balance and resilience.

In summary, the autonomic nervous system is crucial to maintain homeostasis and respond to stress. By regulating involuntary physiological processes through their sympathetic, parasympathetic and enteric components, the ANS ensures that the body works optimally in rest and stressful situations. The ability to adapt to stress and maintain internal processes in balance can significantly influence the physical and mental health of a teenager. Therefore, promoting an understanding of ANS functions and the importance of maintaining their balance could lead us to healthier lives and improving well-being.

Citations:

de Zambotti, Massimiliano, Harold Javitz, Peter L. Franzen, Ty Brumback, Duncan B. Clark, Ian M. Colrain, and Fiona C. Baker. Sex-and age-dependent differences in autonomic nervous system functioning in adolescents. Journal of Adolescent Health 62, no. 2 (2018): 184-190. Https://www.sciencedirect.com/science/article/pii/S1054139X17304743

Tonhajzerová, Ingrid, Kamil Javorka, Michal Javorka, and Mária Petrášková. Cardiovascular autonomic nervous system tests: reference values in young people (15–19 years) and influence of age and gender. Clinical Physiology and functional imaging 22, no. 6 (2002): 398-403. Https://onlinelibrary.wiley.com/doi/abs/10.1046/j.1475-097X.2002.00449.x

Faulkner, Melissa Spezia, Donna Hathaway, and Betsy Tolley. Cardiovascular autonomic function in healthy adolescents. Heart & lung 32, no. 1 (2003): 10-22. Https://www.sciencedirect.com/science/article/pii/S0147956302702069

Harteveld, Lisette M. Ineke Nederend, Arend DJ Ten Harkel, Nienke M. Schutte, Susanne R. De Rooij, Tanja GM Vrijkotte, Helena Oldenhof et al. Maturation of the cardiac autonomic nervous system activity in children and adolescents. Journal of the American Heart Association 10, no. 4 (2021): e017405. Https://www.ahajournals.org/doi/abs/10.1161/JAHA.120.017405

From Cells to Selves: The Essential Journey of Cellular Differentiation in Human Development

As the embryo grows, cells begin to differentiate, meaning they become specialized for specific roles—some become nerve cells, others become skin, muscles, or organs. This is directed by:

Gene expression: Even though every cell has the same DNA, only certain genes are turned on in each cell, based on its location and signals from surrounding cells.

Chemical signals: Cells release and respond to proteins and hormones that help determine their fate.

Positional information: Cells "know" where they are in the developing body, which helps them organize into body structures.

We view these systems not as random or chaotic but as evidence of intelligent design—a Creator who programmed this incredibly complex process with precision.

Formation of Tissues and Organs – Building the Body
Groups of specialized cells form tissues (like muscle or skin), which then form organs (like the heart or brain). These develop in a coordinated way over time, following a timetable laid out in the genetic code.

By about 8 weeks, the basic body plan is established. By the time of birth, the baby is a fully functioning human being, with trillions of differentiated cells—all starting from one fertilized egg.

Cell differentiation is an astonishing process that demonstrates both the complexity of life and the wisdom of the Creator. While we can study and describe these steps scientifically, the underlying order and purpose reflect a divine origin. As creation scientists, we see biology not as a series of accidents, but as the unfolding of a designed plan written into our very cells.

Cell differentiation is a vital component of embryonic development, characterized by the transformation of stem cells not specified in specific types of cells that in the end make up the various tissues and specialized organs of the human body. This complex journey begins with a single fertilized egg, or zygote, which undergoes rapid divisions, culminating in the formation of blastocyst. The blastocyst is an empty sphere consisting of an external layer of cells, known as a trophoblast and an internal cellular mass, which contains multi -canor stem cells. These multiple cells have the considerable ability to differentiate themselves in any type of cell inside the body, ranging from neurons and muscle cells to blood cells and epithelial cells (Wobus & boheler, 2005).

During the differentiation process, various signals guide these stem cells on their path to specialization. One of the primary mechanisms provides for the expression of specific genes that dictate the identity and function of a cell. Factors such as growth factors, hormones and even interactions with nearby cells contribute to the signaling environment that influences differentiation. For example, during the development of the nervous system, some proteins report the precursor cells to become neurons, while others transform them into gliali cells that support and protect neurons. This intricate network of signaling paths guarantees that cells not only adopt their unique roles, but also interact appropriately with the surrounding cells to form functional tissues.

The meaning of cell differentiation extends beyond simple development; It is essential for the establishment of structures and systems organized within the body. Each specialized fabric contributes to a unique function indispensable for general health and homeostasis. For example, in the circulatory system, red blood cells have the task of transporting oxygen, while white blood cells are essential for the immune response. Likewise, differentiated neurons form complex networks that facilitate communication between different parts of the body. Without the differentiation process, the intricate systems required for human life would not be able to form, with consequent lack of organization and functionality through cellular and systemic levels (Keller, 2005).

In addition, the understanding of cell differentiation opens new paths in the field of regenerative medicine. The researchers are exploring how to exploit the potential of stem cells and manipulate the differentiation paths to repair or replace damaged tissues and organs. Techniques such as direct differentiation, in which stem cells are influenced to become specific types of cells in the laboratory, have shown the promise of treatment of a series of conditions from heart disease to spinal cord lesions (Murry & Keller, 2008). By grasping the principles at the basis of differentiation, scientists can develop innovative therapies that one day can restore the function to wounded or sick fabrics.

In summary, cell differentiation is a fundamental process that allows the formation of the fabrics and specialized organs required for human existence. It is a finely tuned and regulated process that guarantees the correct development and functioning of body systems. Through the lens of cell differentiation, not only do we get insights on the wonders of human development, but we also position ourselves at the forefront of the exciting medical progress that promise to change their lives. The meaning of cell differentiation extends beyond the mere formation of distinct cell types; It also plays a critical role in the way these cells are organized into tissues and finally form complex organs.

The differentiation process allows non-specialized stew cells to evolve into various specialized cells, each with an essential structure and function essential for the overall operation of the body. For example, the differentiation of embryonic stem cells in cardiomyocytes, heart muscle cells is crucial for the development of a functional circulatory system (Boheler et al. 2002). The heart, being a vital organ, has the precise formation of these differentiated cells to rhythmically and pump blood throughout the body.

In addition to the creation of specific cell types, the organized arrangement of cells during the early stages of organ development ensures that these structures can perform their specialized functions effectively. The concept of morphogenesis, which refers to the biological processes that make an organism develop its shape, is strongly influenced by time and pattern of cell differentiation. Research indicates that when cells differ, they not only specialize, but also communicate, contributing to the proper structural arrangement for organ functionality (Moore et al. 2011). For example, in the developing brain, different types of neurons and glial cells derive from common progenitor cells, and their respective positions and relationships critically affect circuits and neural function.

In addition, interactions between various cell types are fundamental to the establishment of tissue architecture. Cells do not exist in isolation; They form networks, sending signals and sharing clues that guide tissue formation. This communication is vital to ensure that the organs develop with the correct structure and can operate as intended. Interruptions in these processes can lead to development disorders, which are increasingly recognized as critical areas for research. Understanding cellular differentiation, researchers obtain information on the complexities of developmental pathologies, paving the way for prospective therapeutic interventions (Kiessling & Anderson, 2003). For example, studies on congenital heart defects provide information on how interruptions in cardiomyocyte differentiation can lead to significant health problems, thus informing clinical approaches to treatment.

In short, cell differentiation is fundamental not only for the creation of specialized cells, but also for the subsequent organization of these cells in tissues and organs. The accuracy of this process is essential for the effective formation of systems that maintain biological functions. The intricate network of communication and interaction between differentiated cells supports the delicate balance necessary for the proper development of the organs, highlighting the importance of continuous research on these remarkable biological processes. Ultimately, the implications of cell differentiation go beyond the immediate scope of embryonic development and reach significant ethical considerations relating to embryonic research. In this context, the debate that surrounds the human embryonic stem cells has aroused greater attention, pushing essential investigations on the moral dimensions associated with the manipulation of life to its first stadiums (Holland et al. 2001). Supporters of research on embryonic stem cells underline its potential to produce revolutionary progress in medical treatments for debilitating conditions, including Alzheimer's disease and Parkinson's disease. These conditions, which derive mainly from neuronal degeneration and dysfunction, underline the need to understand cell differentiation processes. The embryonic stem cells have the considerable ability to differentiate themselves in various types of cells, thus offering a unique opportunity to understand and finally mitigate the effects of these diseases through targeted therapies (Dolin, 2009).

In addition, the research that examines the regulation of genes such as PCDH19 in neural progenitor cells illustrates how specific differentiation paths can affect neurological health. PCDH19 is implicated in various neurological development disorders, thus connecting the differentiation models to the various conditions that influence cognitive and motor functions (Homan et al. 2018). Studies that deepen the mechanisms that regulate cell differentiation provide insights on how aberrations can lead to pathological states, thus strengthening the relevance of this research both in therapeutic contexts and on our understanding of human health.

In addition, the intersection of the research and ethics of cell differentiation invites a wider reflection on our social responsibilities in scientific exploration. As we acquire knowledge of differentiation and its implications for human health, it is essential to face these progress with a weighted consideration of the moral implications involved. The potential for manipulating embryonic stem cells for therapeutic use raises significant questions about the moral state of the embryo and on the ethical limits of research that involves human life. These discussions are fundamental while the company is preparing with the opportunities and challenges presented by biomedicine.

The navigation of the complex ethical landscape requires a commitment not only with scientific progress but also the public discourse on the values that should guide our approach to research and application in the field of cell differentiation. Understanding these dynamics allows a more nuanced dialogue that contains both the scientific possibilities and the ethical paintings and the society must consider. By clarifying the critical role that differentiation plays within embryonic development, we can appreciate the complexity of human creation, also recognizing the urgency of responsibly tackling the ethical considerations that accompany these progress in science (Trounson, 2006). The forward path requires a collaborative effort between scientists, ethics, political and the largest community to ensure that the promise of cell differentiation can be exploited in ways that align with our shared moral beliefs and commitments with human dignity.

Citations:

Keller, Gordon. Embryonic stem cell differentiation: emergence of a new era in biology and medicine. Genes & development 19, no. 10 (2005): 1129-1155. Https://genesdev.cshlp.org/content/19/10/1129.short

Wobus, Anna M. and Kenneth R. Boheler. Embryonic stem cells: prospects for developmental biology and cell therapy. Physiological reviews 85, no. 2 (2005): 635-678. Https://journals.physiology.org/doi/abs/10.1152/physrev.00054.2003

Trounson, Alan. The production and directed differentiation of human embryonic stem cells. Endocrine reviews 27, no. 2 (2006): 208-219. Https://academic.oup.com/edrv/article-abstract/27/2/208/2355269

Murry, Charles E. and Gordon Keller. Differentiation of embryonic stem cells to clinically relevant populations: lessons from embryonic development. Cell 132, no. 4 (2008): 661-680. Https://www.cell.com/molecular-cell/fulltext/S009286740800216X

Holland, Suzanne, Karen Lebacqz, and Laurie Zoloth, eds. The human embryonic stem cell debate: Science, ethics, and public policy. Vol. 3. Mit Press, 2001. Https://books.google.com/

The Law of Biogenesis

What is the Law of Biogenesis?

The law of biogenesis states that life only comes from pre-existing life. This principle was solidified through the work of scientists like Louis Pasteur in the 19th century. His experiments disproved the ancient idea of spontaneous generation, which claimed that life could arise from non-living matter (e.g. maggots from rotting meat or microbes from broth).

Pasteur's experiments showed that when sterile, nutrient-rich broth was protected from airborne microbes, it remained lifeless—no life arose unless living organisms were introduced. He famously declared, "Life is a germ and a germ is life."

What Does This Mean for Creation Science?

From a creationist perspective, the law of biogenesis confirms what Scripture has said from the beginning: God is the source of life. In Genesis 1, we read that God created every living thing "according to its kind." Life did not emerge randomly from non-life; it was purposefully created by an intelligent Designer.

This principle also stands in direct contrast to naturalistic evolutionary theories, which assert that life arose from non-living chemicals in some primordial soup billions of years ago. But no scientific observation has ever shown non-living matter producing life on its own. Every known instance of life's origin comes from another living source—exactly what biogenesis affirms.

Key Implications:
Observed science supports biogenesis: in labs, in nature, and in human experience.

Abiogenesis (life from non-life), required by evolution, has never been observed, tested, or repeated—violating the very definition of science.

The law of biogenesis fits a biblical model of life being created by a living God and reproducing "after its kind."

The law of biogenesis stands as both a scientific principle and a powerful affirmation of creation. Life does not emerge from lifeless chemicals by accident. It comes from life—ultimately, from the Author of life Himself (Acts 3:15).

Creation Science Critique of Neo-Darwinism

1. Lack of Evidence for Truly Beneficial Mutations

While Neo-Darwinism depends on random mutations to create new information, most observed mutations are neutral or harmful, not creative.

We argue: "Mutations can break systems, but rarely build them."

2. Information Theory Objection

Adding meaningful, functional genetic information (e.g., to evolve a new organ) requires more than random change.

We assert that information requires an intelligent source, not chance.

3. Irreducible Complexity

Systems like the bacterial flagellum or blood clotting cascade are irreducibly complex—meaning all parts must exist simultaneously to function.

This challenges the step-by-step evolution Neo-Darwinism proposes.

4. Fossil Record Discrepancies

Neo-Darwinism predicts gradual transitions, but the fossil record often shows sudden appearance and stasis (as Stephen Jay Gould admitted with punctuated equilibrium).

We view this as evidence for created kinds rather than slow evolution.

5. No Observation of Macro-Evolution

We differentiate between microevolution (variation within a kind) and macroevolution (change between kinds).

While microevolution is observable and accepted, they argue macroevolution remains theoretical and unproven.

6. Philosophical Bias Toward Naturalism

Neo-Darwinism, they argue, excludes any intelligent cause by definition, not by evidence.

We see this as a metaphysical commitment to materialism, not objective science.

"Cells of Wonder: Unveiling the Intricate Design and Intelligent Creation in the Microcosm of Human Life"

Human cells embody the peak of biological complexity, acting as fundamental units of life that embody a significant series of specialized structures and functions. Each human cell operates with precision and organization that reflects a sophisticated design, well beyond simple random mutations or random evolutionary processes. At the center of cell functionality is the nucleus, an organelle that contains the genetic material that orchestrates cell activities through expression and gene regulation. This intricate control system determines the various roles that cells play, from muscle contraction to hormonal production, underlining the complexity that defines multicellular life (Haught, 2018).

To complement the nucleus is the mitochondrion, often nicknamed the cell power plant, which meticulously converts nutrients into adenosine triphosphate (ATP), the essential energy currency for a myriad of cellular processes. The structure of the mitochondria, with a double membrane and its DNA set, suggests not only efficiency in the production of energy but also an evolutionary lineage that speaks to symbiotic relationships, improving the topic for intelligent design rather than the Serendipity of evolution.

In addition, ribosomes play a critical role in protein synthesis, translating genetic codes into functional proteins necessary for structure and cell function. Their ability to assemble amino acids into polypeptides with considerable precision further confirms the complexity inherent in human cells. Upon a more careful examination, it is clear that each cellular component, from the endoplasmic reticulum to the Golgi apparatus, has unparalleled coordination and specialization that collectively make the cell an extraordinary living machine (Haught, 2018).

Approaching the conversation of complexity from an evolutionary perspective introduces a nuanced discourse surrounding intelligent design. As the philosopher Daniel Dennett discusses, the mechanisms of biological evolution can coexist with an appreciation of the elaborate complexities observed within cellular structures. It posits that while natural selection can explain the adaptation of traits over time, the emerging complexity of life—particularly at the cellular level—raises deep questions regarding the sources of that complexity (Dennett, 2017).

The ongoing debate hinges on whether these multilayer cellular systems can derive exclusively from random processes or if they indicate an orchestrated origin. Evidence of intricate interdependencies between cellular components; for example, the finely tuned interactions between signaling pathways and metabolic pathways suggest a deliberate design. The ability of human cells to self-replicate, respond to environmental stimuli, and maintain homeostasis reflects a dynamic system that is not simply functional but also resilient, suggesting a general intelligence behind the architecture of life.

This discussion lays the foundations for further dissection of the various cellular components and their roles in demonstrating intricate design. Analyzing the structural and functional complexities of human cells, parallels can be drawn to intelligent creation, strengthening the argument that this sophisticated microcosmic life must be the product of intentional design, deserving of deep exploration and appreciation. The functional sophistication of human cells illustrates an extraordinary complexity that suggests a deliberate conception, essential for maintaining life as we know it. Each specific type of cell shows tailor-made characteristics and capacities that are perfectly aligned with the crucial roles they play in the physiological symphony of the human body. Stem cells, for example, serve as the basis for regeneration and cell differentiation, having the unique capacity to develop into any type of cell required for the repair and growth of tissues. This remarkable feature highlights their importance in maintaining the integrity and adaptability of the body. According to Rescher (2020), the complexity inherent in stem cells raises deep philosophical questions about the nature of life and the mechanisms underlying biological organization, suggesting that the complex architecture of these cells is proof of intelligent conception.

Likewise, neurons, as specialized cells in the nervous system, have distinct structures and functionalities that are integral to information processing and communication throughout the body. Their long extensions, called axons and dendrites, create a vast network that facilitates the rapid transmission of electrical impulses, allowing humans to react to stimuli and engage in complex cognitive processes. The remarkable precision and efficiency of neural connections underline the sophistication of cellular design and imply an underlying intelligence in their assembly and operation. The capacity of neurons to form synapses, to strengthen experiences, and to adapt over time—known as neuroplasticity—illustrates the dynamic and determined nature of cellular functions, aligning with philosophical inquiries into the intentionality underlying complex systems.

Muscle cells, or myocytes, also demonstrate the specificity of cellular design adapted to functional excellence. These cells have unique structural adaptations such as striations and multinucleation, which allow powerful contractions essential for movement, posture, and circulation. The coordinated action of muscle fibers is not simply a mechanical process; it is an orchestrated event that requires precise molecular signals and energy use. The capacity of muscle cells to contract and relax in a regulated manner illustrates a level of extraordinary complexity similar to that found in advanced engineering concepts, echoing the metaphor presented by Bengisu and Ferrara (2018) concerning intelligent materials that dynamically respond to environmental stimuli. Just as intelligent materials are designed to adapt and exhibit intelligent behavior, muscle cells also illustrate adaptability in their biological context.

Overall, the multifaceted features of stem cells, neurons, and muscle cells illustrate a broader story of cell complexity, reflecting a level of intentionality and precision suggesting intelligent creation. The interconnected roles that these cells play in the maintenance of life reveal an astonishingly complex microcosmic environment, where each component contributes to the greater functioning of the organism as a whole. The philosophical ramifications of such complexity require a re-examination of traditional concepts of design in biology, arguing for the recognition of intelligence manifested by the delicate orchestration of cellular functions. The intricate design and functionality of human cells offer convincing evidence of the concept of intelligent design, particularly when the complexity integrated into their structures and processes is considered. As described by Pobiner (2016), the discussion surrounding the evolution and understanding of biological complexity requires a paradigm shift in how we perceive educational methodologies in biology. While evolutionary theory provides a framework to understand the processes that have led to this complexity, the question of how such sophistication has emerged remains significant. This invites consideration that the biological systems within human cells could be manifestations of intelligent design.

Pobiner (2016) argues that the integration of evolutionary perspectives in education should not rule out the notion of intelligence behind biological complexity. On the other hand, educators and researchers must encourage deepening the symbiotic relationship between evolutionary theory and the recognition of the intricate organization of biological entities. The human cell is equipped with numerous components, including the organelles that perform specialized functions, intricate signaling pathways that maintain homeostasis, and complex mechanisms such as DNA replication and protein synthesis. Each of these elements works harmoniously, demonstrating an organized complexity reminiscent of human engineering systems. Taking these elements into account through the lens of intelligent design allows for a broader appreciation of the potential purpose behind biological function, placing cells not simply as products of random processes, but as results of directed and reflective creation.

In addition, contemporary advances in artificial intelligence (AI) support the argument for intelligent design by paralleling biological complexity with engineering systems that exhibit problem-solving capabilities. Quan et al. (2019) explore the implementation of design principles in AI systems that mimic the biological processes observed in human cells. These systems often incorporate mechanisms for self-regulation, adaptation to changing environmental conditions, and the optimization of tasks, qualities that reflect the functions of cellular processes such as feedback loops in metabolic pathways and the adaptability of immune responses. This technology, influenced by biological design, also illustrates that human engineers are often inspired by the sophistication of nature, hinting at an underlying plan or intelligence inherent in these biological structures.

Shen et al. (2019) take this discussion one step further by highlighting parallels between cellular functions and engineering systems in terms of resilience and efficiency. For example, just as engineering networks are designed to resist failures and adapt to inputs, human cells exhibit remarkable resilience in various conditions. The ability of cells to reprogram and adapt their metabolic pathways in response to stressors is similar to the principles of redundant system designs in artificial technologies, which are based on avoiding individual failure points. This reflexive consideration leads to the conclusion that a deliberate intelligence may be at stake in the structural and functional complexity of the cells, a design that transcends mere chance.

The rich tapestry of human cell life, woven with the threads of complexity and functionality, highlights an intentionality that invites contemplation of intelligent design. When building upon the ideas derived from evolutionary principles and the technological advances of the avant-garde, this research opens pathways for a deeper understanding of the origins of life and the magnificent complexities within us. As we continue to unravel the layers of functionality and cell design, the narrative of intelligent creation becomes an integrative framework through which both scientific research and the appreciation of the wonders of life can coexist. The study of human cells allows us to appreciate the profound complexity that supports our existence, which can be perceived as proof of intelligent design. From the initial formation of the zygote to the tissues and organized organs, the refinement of cellular structures reveals a level of specificity and functionality that suggests an intelligent creator. Essential organelles such as mitochondria, with their role in energy production, and the endoplasmic reticulum, fundamental in protein synthesis and lipid metabolism, illustrate significant interaction systems that contribute to the cellularism and overall health of the body. In addition, the highly regulated processes of communication and cell division reflect an intelligence that is not only functional but also anticipates the different challenges faced by living organisms.

As we deepen our understanding of the intricate architecture of human cells, it becomes increasingly evident that this complexity transcends simple possibility or evolutionary adaptation. Finely tuned genetic mechanisms regulate not only survival but also a complex network of interactions within and between cells. This weighted design paradigm invites a review of the debate surrounding creationism and evolution. Reiss (2014) claims that understanding the complexity of biological systems can inform discussions about the origins of life, potentially filling gaps between apparently opposite points of view.

In contemplating the intersection of cellular complexity and philosophical theories, Wilensky and Rand (2015) highlight the characteristics of complex systems that engage not only in self-organization but also in the emergence of new properties through the interactions of their parts. The human body exemplifies this principle; cellular networks dynamically respond to stimuli, displaying adaptability and maintaining resilience in the face of diseases and environmental challenges. This natural ingenuity reflects a fundamental aspect of computational thought—the ability to devise solutions and innovate through the interrelation of simpler entities—a capacity exemplified by the collaborative functioning of human cells. Shute et al. (2017) point out that the creative processes seen in computational systems parallel biological processes, suggesting that the lessons learned from cellular function can inspire technological progress and innovative applications.

Exploring the multifaceted nature of human cells, it is clear that the complexity observed within these microcosmic lives signifies a design that is intentional, organized, and inevitably suggestive of intelligent creation. The interaction of cellular components, governed by complex regulations and emerging behaviors, not only underlines the wonder of life itself but also provides fertile ground for innovation and interdisciplinary exploration. Recognizing the profound depth of cellular systems and their implications invites dialogue that transcends conventional boundaries, promoting a richer understanding of how life manifests and thrives. The intricate design of human cells, therefore, is not only a testament to the complexity of biological phenomena but also a springboard for future interpretations of existence and progress, filling scientific investigation with philosophical reflection.

Citations

Haught, John F. God after Darwin: A theology of evolution. Routledge, 2018. https://www.taylorfrancis.com/books/mono/10.4324/9780429500169/god-darwin-john-haught

Quan, Steven Jige, James Park, Athanassios Economou, and Sugie Lee. Artificial intelligence-aided design: Smart design for sustainable city development. Environment and Planning B: Urban Analytics and City Science 46, no. 8 (2019): 1581-1599. https://journals.sagepub.com/doi/abs/10.1177/2399808319867946

Bengisu, Murat, and Marinella Ferrara. Materials that move: smart materials, intelligent design. Cham: Springer International Publishing, 2018. https://link.springer.com/content/pdf/10.1007/978-3-319-76889-2.pdf

Rescher, Nicholas. Complexity: A philosophical overview. Routledge, 2020. https://www.taylorfrancis.com/books/mono/10.4324/9780429336591/complexity-nicholas-rescher

Pobiner, Briana. Accepting, understanding, teaching, and learning (human) evolution: Obstacles and opportunities. American journal of physical anthropology 159 (2016): 232-274. https://onlinelibrary.wiley.com/doi/abs/10.1002/ajpa.22910

Shen, Weiming, Douglas H. Norrie, and Jean-Paul Barthès. Multi-agent systems for concurrent intelligent design and manufacturing. CRC press, 2019.

Shute, Valerie J., Chen Sun, and Jodi Asbell-Clarke. Demystifying computational thinking. Educational research review 22 (2017): 142-158. https://www.sciencedirect.com/science/article/pii/

S1747938X17300350

Dennett, Daniel C. From bacteria to Bach and back: The evolution of minds. WW Norton & Company, 2017.

Wilensky, Uri, and William Rand. An introduction to agent-based modeling: modeling natural, social, and engineered complex systems with NetLogo. MIT press, 2015.

Reiss, Michael. Creation or Evolution Do we have to choose….?. (2014). https://www.researchgate.net/profile/Michael-Reiss-4/publication/255606072_Creation_or_evolution_-_do_we_have_to_choose/links/0c960536403b452b80000000/Creation-or-evolution-do-we-have-to-choose.pdf

Summary

From the creation science viewpoint, the human cell is not merely a product of random chance or undirected processes, but a masterpiece of intelligent design—an intricate microcosm that reflects purpose, order, and foresight.

1. The Cell: A City of Purposeful Design

The human cell is often compared to a bustling city, with specialized structures (organelles) performing distinct, coordinated tasks. Each cell contains billions of molecules, all working in harmony. This level of organization and interdependence points to a blueprint, much like a city requires architects and planners.

Nucleus: The command center, housing DNA—the cell's information library. DNA contains instructions for building and maintaining the cell, encoded in a language of four chemical "letters." The sheer volume and accuracy of this information (over 3 billion base pairs in humans) is staggering, and its error-checking and repair mechanisms suggest foresight and intentionality.
Mitochondria: The power plants, converting nutrients into usable energy (ATP) through complex biochemical pathways. The efficiency and regulation of this process are so precise that even minor disruptions can be catastrophic.

Endoplasmic Reticulum & Golgi Apparatus: The manufacturing and shipping centers, assembling proteins and lipids, then packaging and sending them to their destinations. The specificity of protein folding and transport is so exact that it defies the odds of random assembly.

2. Molecular Machines: Evidence of Engineering
Within the cell, we find molecular machines—complex protein structures that perform specific tasks. Examples include:

ATP Synthase: A rotary motor that produces ATP, the cell's energy currency. It spins at thousands of revolutions per minute, powered by a flow of protons, and is so efficient that it inspired nanotechnology research.

Kinesin and Dynein: Motor proteins that "walk" along microtubules, carrying cargo to precise locations. Their movement is coordinated, purposeful, and highly regulated.
These machines are irreducibly complex; they require all their parts to function. Remove one component, and the machine fails. This suggests that such systems could not arise stepwise through gradual mutations, but rather point to an intelligent cause.

3. Information and Communication: The Language of Life
Cells communicate using chemical signals, responding to their environment and coordinating with other cells. The genetic code itself is a language, with syntax and semantics, capable of storing and transmitting vast amounts of information. In information science, codes and languages always originate from intelligent minds, not random processes.

4. Self-Repair and Reproduction: Hallmarks of Design
Cells can repair their own DNA, replace damaged components, and even self-replicate. The process of mitosis (cell division) is tightly regulated, ensuring accurate copying and distribution of genetic material. Such self-maintenance and reproduction are hallmarks of systems designed for longevity and adaptability.

5. Fine-Tuning and Optimal Function

Many cellular processes operate at the edge of physical and chemical limits, optimized for efficiency and reliability. Enzymes catalyze reactions at rates millions of times faster than would occur naturally. The cell membrane is selectively permeable, maintaining homeostasis with remarkable precision.

Conclusion: The Signature of a Designer

From a creation scientist's perspective, the human cell is a marvel of microcosm—an intricate, purposeful, and information-rich system that bears the unmistakable hallmarks of intelligent design. Its complexity, interdependence, and functionality far exceed what undirected processes could plausibly achieve. Instead, the cell points to a Creator who engineered life with wisdom, foresight, and care.

BIBLIOGRAPHY

Books

Algorithms and the Neo-Darwinian Theory of Evolution
- Marcel P. Schutzenberger - the University of Paris

Behavioral Genes Why We Do What We Do— and How to Change
James D. Baird, Ph.D.

Biology AP Test Prep - Phillip Pack - CliffsNotes

Biology: Explaining the Cell Cell Structure and Organelles, Cell Specialisation and Function- Jonathan Clark -

The Body Book - David Bodanis - Little Brown

Body By Design - Fearfully and Wonderfully Made - Dr. Alan L. Gillen - Master Books

A Civic Biology by George Hunter

Chance and Necessity - Jacques Monod

By Design - Evidence for nature's Intelligent Designer - The God of the Bible - Dr. Jonathan Sarfati - Creation Book Publishers

By Design - Science and the Search for God - Larry Witham - Encounter Books

The Collapse of Evolution - Scott Huse - Baker Books

Answers from the Bible & Science - Richard Myer - www.booksurge.com

CREATION BASICS & BEYOND An In-Depth Look at Science, Origins, and Evolution - Institute for Creation Research

Creation Facts of Life - How Real Science Reveals the Hand of God - Gary Parker PhD - Master Books

Darwin's Black Box - Michael Behe - Free Press

Darwin's Demise - Joe White and Nicholas Comninellis - Master Books

Darwin's Mistake - Tom Bethell - Harper's

Darwin Retried: an Appeal to Reason - Norman Macbeth - Harvard Common Press

Darwin on Trial - Phillip Johnson - InterVarsity Press

Darwin Was Wrong - A Study in Probabilities. - L.L. Cohen - New Research Publications, Inc.

Darwinism Under the Microscope - James Gills M.D. and Tom Woodward, PhD - Charisma House

Essential Cell Biology - Garland Science - Taylor and Francis Group

Evolution of Living Organisms - Pierre-Paul Grassé,

Fundamentals of Anatomy & Physiology - Charles M. Seiger - Prentice Hall

The Edge of Evolution - The Search for the Limits of Darwinism - Michael Behe - Simon & Schuster

THE LAW OF THE EMPEROR'S NEW CLOTHES VOL. 2 HAND OF GOD Why Macroevolution is a Scientific Fact When it's Really a Lie - PETER-BRIAN ANDERSSON - Good News Publishers.
Evolution A Theory in Crisis - Michael Denton - Adler & Adler

Evolution is STUPID - John Verderme - GioiaMia

Evolution or Creation? A Comparison of the Arguments Third Edition - Albert J. DeBenedictis - Xlibris

Exploring the Evidence for Creation - Henry Morris III - Institute For Creation Research

Evolution from Space - Hoyle and Wickramasinghe, - Aldine House

How the Body Works - Dr. Peter Abrahams - Amber Books Ltd.

Icons of Evolution - Science or Myth? - Jonathan Wells - Regnery Publishing, Inc.

Human Body - Ann Baggaley - DK

Human Body - A comprehensive Guide to the Wonders of the Body - Quantum Publishing

Icons of Evolution - Jonathan Wells - Regnery

In Six Days - Why 50 Scientists Choose to Believe Creation - John Ashton PhD - Master Books

In the Beginning was Information A Scientist Explains the Incredible Designs in Nature - Dr. Werner Gitt - Master Books.

In the beginning - Walter Brown

Interpreting Origins Science - David Rosevear - Creation Science Movement.

The Lie: Evolution - Ken Ham - Master Books

Life Itself -- Its Origin and Nature, Futura - Francis Crick,.

Mere Creation - Science, Faith and Intelligent Design - William De
mbski - InterVarsity Press

Metabolism Basics: A Walkthrough Guide for Fermentation and Cellular Respiration - Jamie Jacobs - Walkthrough Basics

Microbiology - Lansing Prescott, John Harley, Donald Klien - McGraw-Hill

Microbiology - Demystified - Tom Betsy D.C. and Jim Keogh R.N. - Mcgraw Hill

Molecular Cell Biology - (multiple authors) - W. H Freeman and Company

More Than Meets the Eye: Fascinating Glimpses of God's Power and Design - , Richard Swenson - NavPress

The Mysterious Epigenome: What Lies Beyond DNA -Thomas E. Woodward and James P. Gills - Kregel Publications,.

The Myth of Junk DNA - Jonathan Wells - Discovery Institute Press

The Myth of the Flat Earth - Jeffrey B Russell - American Scientific Affiliation.

Nutrient Power - William J. Walsh, PhD - Skyhorse Publishing.

OBESITY GENES - James D. Baird, PhD - HWL. Inc.

Our Straight-Forward Origins in everyday Language - Knox Chilumbu

Principles of Anatomy and Physiology - Tenth Edition - Tortora Grabowski - Wiley

The Politically Incorrect Guide to Darwinism and Intelligent Design - Jonathan Wells - Regnery

Refuting Evolution - Jonathan Sarfati PhD - Master Books

Resynthesizing Evolutionary and Developmental Biology Scott Gilbert, John Opitz, and Rudolf Raff

THE SEEDS of life - Joseph Ancion -

The Seven Daughters of Eve - Bryan Sykes

Signature in the Cell - Stephen Meyer PhD - Harper One

Structure and Function of The Body - Gary Thibodeau, PhD and Kevin Patton, PhD - Mosby
What Darwin Didn't Know - Geoffrey Simmons, M.D. - Harvest House Publishers

Uncommon Dissent - Intellectuals Who Find Darwinism Unconvincing. - William Dembski Editor - ISI Books

What Your Atheist Professor Doesn't Know (But Should) An exploration of the implications of modern science, archaeology, history and philosophy - Stephen Joseph Williams - RFH

The Wonders of Life, (trans. J. McCabe) p. 111. Haeckel (1905),

Articles:

Body plan: from genes to embryo. ('homeo box' segment of DNA thought to direct body development in wide variety of animals)." Science News. Science Service, Inc. 1985.

"*Confessions of a Professed Atheist,*" Report: Perspective on the News, Aldous Huxley. Vol. 3, June 1966, p. 19

Correspondence to Nature 401(6752):423, 30 Sept. 1999; - S.C. Todd

Keeping an Eye on Evolution: David Berlinski *Richard Dawkins, a relentless Darwinian spear carrier, trips over Mount Improbable.* The Globe & Mail, November 2, 1996, p. D10

Genome at 10: Faster, cheaper... worse - Michael Le Page - New Scientist

Mutations mean death and disease - Kelly Hollowell, J.D. Ph.D.

The Natural Limits to Biological Change - Dr. Ray Bohlin

What is cancer? by Peter Crosta for Medical News Today in September 2008. © MediLexicon International Ltd

Gould, Stephen J. in Lecture at Hobart and William Smith College. February 14, 1980

Origins Research 16(1):9, 1994. - William Provine

Journal of American Chemical Society, Vol. 97

What are Stem Cells? - National Stem Cell Foundation

Training Companies:

Wyzant.com - contract tutors

Biology - Concord Consortium, Inc.

University of Utah - Health Sciences - *Learning Genetics.*

Video Series:

Basic Immunology - Nuts and Bolts of the Immune System - University of California - Anthony DeFranco, MD

Body of Evidence Series - 8 lectures - Dr. David Menton - Answers In Genesis

Case for the Creator - Lee Strobel - Illustra Media

Creation and the Christian Faith - Answers in Genesis

Darwin's Dilemma - Illustra Media

Expelled - No Intelligence Allowed - Ben Stein

Fearfully and Wonderfully Made - Dr. David Menton - Answers in Genesis

God of Wonders - Exploring the Wonders of Creation, Conscience and the Glory of God- Eternal Productions

Human Physiology The Immune System - Dr. Ketchum - University of Oklahoma

Immunology - University of California Irvine - David Fruman, PhD.

Introduction to Clinical Immunology - University of California San Francisco - Katherine Gunding, Md. MPH

Necessity of the Immune System - Stanford University - Carl Naumann - Professor

Understanding the Brain 36 Lectures Professor Jeanette Norden Ph.D. Vanderbilt University - The Great Courses

Understanding the Human Body: An Introduction to Anatomy and Physiology - 32 Lectures - Dr. Anthony A. Goodman M.D. - Montana State University - The Great Courses

Why We don't Heal - Dr. Anthony A. Goodman M.D. - Montana State University - The Great Courses

Wonder of the Cell - Dr. Georgia Purdom- Answers In Genesis

Various clips from UCLA, Stanford and other major universities

Made in the USA
Monee, IL
04 November 2026

67380506R00193